To Dave

CW00505210

Best wishes

21. 7. 2016.

A good education at Kirkham Grammar School preceded an eclectic mix of occupations and interests for the author, and prepared him for most of the challenges he has met.

The age of sixty years saw him concentrate on his out-of-work interests of natural history, particularly in the Inner and Outer Hebrides. The distant islands of St. Kilda in the Hebrides sparked an interest in the social history of the island, an interest he brought home with him and applied to his birth-town of Preston.

Then he found Moses Holden, and his life changed!

MOSES HOLDEN 1777 – 1864

A Freeman of Preston, Astronomer and Self-taught Genius - A Man who was Never Eclipsed.

Stephen R. Halliwell

Moses Holden
1777 – 1864

A Freeman of Preston, Astronomer and Self-taught Genius - A Man who was Never Eclipsed.

Vanguard Press

VANGUARD PAPERBACK

© Copyright 2016
Stephen R. Halliwell

The right of Stephen R. Halliwell to be identified as author of
this work has been asserted by him in accordance with the
Copyright, Designs and Patents Act 1988.

All Rights Reserved

No reproduction, copy or transmission of this publication
may be made without written permission.
No paragraph of this publication may be reproduced,
copied or transmitted save with the written permission of the publisher,
or in accordance with the provisions
of the Copyright Act 1956 (as amended).

Any person who commits any unauthorised act in relation to
this publication may be liable to criminal
prosecution and civil claims for damages.

A CIP catalogue record for this title is
available from the British Library.

ISBN 978 17845 115 2

*Vanguard Press is an imprint of
Pegasus Elliot MacKenzie Publishers Ltd.*
www.pegasuspublishers.com

First published in 2016

Vanguard Press
Sheraton House Castle Park
Cambridge England

Printed and bound in Great Britain

Dedication

In a life that is seemingly more and more difficult and uncertain for our leaders of tomorrow, I would like to dedicate this book to my precious grandchildren, Oscar, and the twins, Phoebe and Eliza, in the hope that someday they may draw inspiration from it, and like my subject, believe that nothing is impossible with the right application, dedication, and above all, hard work.

Acknowledgements

I am particularly grateful to the Lancashire Record Office Archives for allowing me the use of a copy of Moses Holden's Certificate as Freeman of the Borough of my birthplace, Preston. I am also thankful that my wife, Rita, has given me the time and encouragement to follow my research into a fascinating life and character to a natural conclusion.

Furthermore, I am indebted to Professor Derek Ward-Thompson, Director of the Jeremiah Horrocks Institute at UCLan, Preston, for his interest and encouragement in my project.

Finally, I would like to thank Patrick B. Holden, the three-times great grandson of Moses Holden, who, on behalf of various generations of his extant family, has given me access to material that can only ever be available from the family.

Moses Holden

Chapter One

Introduction, and the years before 1800

My research into early nineteenth century learned societies connected to science and literary matters in Preston seem to be exposing the same names. One of the more frequent of those individual's names was that of Moses Holden, a local astronomer and lecturer, and, it was to transpire, much more. The more that I learned of the man, a feeling of needing to learn even more was kindled. My searches to find more about his contemporaries and associates, his family and his varied endeavours throughout his life, rarely resulted in disappointment, and yet his name seemed to mean little or nothing at all to those with whom I spoke.

Slowly but surely, slivers of information began to emerge about Moses Holden, his commitment to his religion, his passion for astronomy, his skill as a self-taught mechanic, and his talent and reputation as a travelling lecturer. Then, when I was least expecting it, I discovered an article in the *Preston Herald* a century later. On the 24th April 1942, J. H. Spencer, a reporter with that newspaper wrote, 'A Forgotten Freeman of Preston – The Hand-loom weaver who became a Famous Astronomer.'

In the article he referred to having been shown a collection of documents by the borough librarian, Joseph Pomfret that had been deposited at the library in 1933 by Miss Mary B. Moorcroft, a great grand-daughter of Moses, on behalf of herself and her two sisters, Martha and Margery. All three were spinsters. Mary, incidentally, lived until 1991, by which time she had reached her 102nd year.

The documents that Spencer was given access to included the original certificate from when Moses was elected a Freeman of Preston in 1834 and a letter from Richard Palmer, the Town Clerk that accompanied it. A 'long and interesting' letter from William Rogerson at the Greenwich Observatory dated 1836. Moses Holden's Celestial Atlas, or Map of the Visible Heavens, 4th edition, and three small memo books containing a journal kept by Moses in 1811 – 1812 recording his experiences whilst a Methodist preacher in the Fylde and Over Wyre district.

Armed with a copy of Spencer's article I asked the library staff if I could have a look at them, but no record of the deposit existed. Yet, in the absence of information to indicate that they had been sent elsewhere, I concluded that they must still be in the Harris complex, and I submitted a report that conveyed that view.

At that time, a new lift shaft was being incorporated at the museum, and the work involved sacrificing a room adjacent to the Reference Library that had hitherto been used as a storeroom. It was necessary to empty the room, and during the course of that work, the person to whom I had addressed my report, discovered a large manila envelope. On each side of the envelope were the words 'Moses Holden,' and documents that probably hadn't seen the light of day for in excess of sixty years had been rediscovered.

Shortly afterwards I became aware that Miss Mary Moorcroft had, in 1933, also made a deposit of some photographs at the Manchester Record Office. Prior to that I'd held out little hope of adding an image of my subject to the word picture I was slowly building, despite knowing that a lithographic print had been made from a portrait of him that was painted by talented Preston artist Charles Hardwick, a man who is probably better known for his books about Preston.

The collection in Manchester Record Office was described in their catalogue as 'Photographs of the Holden Family,' and consisted of 784 thumb-nail images, each with a note in fine handwriting underneath to indicate their content. There was just one, where the indication was given as 'thought to be Moses Holden.' The thought that it may also, not be Moses Holden occurred to me, but I decided to acquire an enlargement of it.

I took the image to show the Keeper of Social History at the Harris Museum, Emma Heslewood, for, as we shall see in Chapter Four, she and local historian John Garlington, had been researching the work of Robert Pateson, an early Preston photographer, and as we'll discover, they believed that Moses may have been connected to him relative to the grinding and polishing of lenses.

Three months later I received a telephone call from Emma, who told me that she had discovered a framed ambrotype photograph in the cellars of the museum that she instantly recognised as being of the same subject. She turned the photograph over, and on the back of the frame was written 'Moses Holden, died 3rd June 1864.' The Manchester image *was* Moses Holden, and a circle had been closed.

Moses Holden has been described by me simply as a genius. The majority of people who are accorded such an accolade have acquired it through their exceptional achievements in their pursuit of one, or occasionally two, areas of expertise, but in my view, this man became a prodigious exception, among many fascinatingly exceptional individuals in our past. Indeed, a multi-faceted genius! Had he any flaws? Probably; in later life he was described as 'difficult to deal with'. Had any of his family any flaws? His eldest son, William Archimedes, was certainly flawed, as we will discover; but are not we all to a greater or lesser extent?

This Bolton-born, Preston resident and ultimate Freeman of the town, was once described as a short, broad-set man, of strong physique and even stronger will, of vigorous intellect, and had such an eclectic range of interests that clearly consumed his nigh eighty-seven year life, yet it is difficult to decide the ideal way to present an account of it. If a thread to his life were to be sought, it might be found in the way he strived, with a strict Wesleyan Methodist upbringing 'to do good' in all the things with which he involved himself. I hope that I am able to demonstrate that over-arching thought as we travel chronologically through his long life from the humblest of beginnings in two Lancashire mill-towns.

In his History of Preston (1837), Peter Whittle, a man who is certain to have known Moses Holden, wrote, *"We full well know, and consider that had Moses Holden been a classical scholar, great additions in the display of science would have been given by him: he most certainly possesses the propensity to constructiveness in a high degree. If to this had been given the intellectual and perceptive faculty of language, the great work in one man would have been more fully developed."* We will see in a later chapter how he could be more than a match with scientists who had far greater qualifications when he attended the meetings of the British Association for the Advancement of Science.

Moses was born in Black Horse Street, Bolton, on the 21st November 1777, the son of handloom weaver Thomas Holden and his wife Joyce. He had a brother, John, who was about two years his senior, and he was succeeded by two further brothers and a sister, James born in 1783, William, whose year of birth is unknown, and Alice, who was born in around 1780. Whether the family was complete prior to its removal to Preston isn't clear, but is relatively unimportant. The thought is that they moved to Preston in about 1882, for in later life Moses talked of being about five years old at the time.

Whilst handloom weavers may not have been the lowest of paid people in the 1770s, Thomas's rapidly growing family must have put an increasing strain on the household budget, and the future life and careers of the children will certainly not have had the benefit of being shaped and formed in the same way that many people achieved the sort of notoriety and success that Moses enjoyed, by their good fortune of being the product of a wealthy family – the silver spoon syndrome.

I hope, through the next several pages to be able to show that Moses Holden achieved what he did through steely determination and gritty self-help. Indeed, Moses probably explained it perfectly at the formation of the Institute for the Diffusion of Knowledge when he said, "The fact is, Sir, I had to make my way solitarily, and take as it were the pickaxe and cut my way through the solid rocks, without assistance or help from anyone; but I laboured at it; yes, I persevered and fainted not." He continued by saying, "While others were sleeping in their beds I was acquiring knowledge. Besides the works on science were costly, I had those to procure alone, and did procure them with much labour and great expense." Moses was describing the advantages offered by the institution that he had been instrumental in creating, and in particular the benefit of spreading the cost of reading material to form a library, which would be of service to those who wished to take advantage of it for many years to come.

Little is known about his four siblings, but for the benefit of my reader I have shown in Appendix One, a three-tier family tree to make the picture clearer.

Two things seem to have conspired when the two eldest boys John and Moses were very young. Firstly, their father read a great deal to them, and among the more popular stories was one about Jeremiah Horrox, the 17th century astronomer, who, when Curate of St. Michael's Church, Hoole, just a few miles from Preston, became the first astronomer to predict and then witness the passage of Venus over the face of the Sun. Horrox became so inspirational to the life of Moses that he provided, in

1826, a memorial plaque that is still displayed in the coincidentally named St. Michael's Church in Horrox's birthplace of Toxteth Park, Liverpool. He may well have been a similar inspiration to his brother John, for family records tell us that he also was an astronomer and mathematician, but he was dead by 1810 at the age of thirty-five. Moses had begun to give lectures by this date, as well as acting as a Methodist missionary, but the extent of John's involvement, if any, is not known.

The other element in the conspiracy also stemmed from their father. In this case it had its roots in Thomas's strict upbringing of his children, with the 18th century version of the 'naughty-step' being the boy's attic bedroom where there was only a sky-light, and the only view was in an upwardly direction. The presumed and unintended outcome of what Moses was in later life to describe as their imprisonment, was to sow a seed in the minds of the two boys about what there was to be learned about the heavens, and it was to be a seed that had a profound effect on them both.

After moving to Preston in around 1782, the family are said to have gone to live 'at an isolated homestead on the outskirts of the town.' Although it is difficult to imagine, it is thought that it may have been at the northern end of Friargate, and although Preston was the first provincial town to have street gas-lighting, it was still thirty years away from its introduction. The night skies would be perfect for the astronomer, and the two boys, who were now separated from their Bolton playmates, would enjoy playing together in the neighbouring fields. It is said that Moses said to his brother, "John, if we knew something about these stars, these mysterious orbs, we should never be lonely again. We should be as happy as when we had our playfellows." Interestingly, it was recorded by Peter Whittle in his History of Preston of 1837, that John responded by saying, "We could never know what they were, for our father would not let us," and this surprising remark was said to be the catalyst that instilled the determination that Moses required "to find out all that there was to be known about the stars."

The two brothers began by gaining knowledge of figures, starting with the four rules of arithmetic, and they strove to understand them. From these, it is said, Moses quickly moved on to logarithms and algebra, both of which he found 'a pleasing study', and by perseverance, he was enabled to pursue his favourite subject. Later in life he was to reflect on what he had achieved and observed that 'It was only by *loving* any particular branch of knowledge, and working assiduously at it, that it could be mastered.' He went on to compare his achievements to the miser, who 'achieved riches through their sheer love of money, looking after it, and keeping hold of it after it had been acquired.'

He gave further hints regarding his by now well-established character at the formation of the Mechanic's Institute in 1828 when he said in the speech to which I earlier referred, that "People must not look for much assistance from others: they must go and plough through it themselves. Chemistry was almost the soul of philosophy, and its study had the power to unfold the best powers of the mind. The study of geometry also tended to give a precision to the mental faculties, and a disposition to correct reasoning and useful enquiry." Touching on his religious views, Moses continued his speech by saying that it had been erroneously said by others that studying such subjects was inimical (or hostile) to religion. Moses' view was that the lustre of Christianity couldn't be tarnished, and education of such things could only serve the mind to judge more seriously and devoutly the wonders of creation. Incidentally, the Institute for the Diffusion of Knowledge would, in most towns and cities have been called the Mechanic's Institute, but it was at Moses' behest that the rather grand alternative name was given to it. We'll read more about that in Chapter Four, but the words contained in this paragraph really do beg a number of questions as to how he managed to achieve what he did. His limited education at Sunday school was restricted to him learning from an old woman how to read the Psalms. Hardly a firm foundation?

However, long before such a distant date as 1828, Moses life was being formed and fashioned by living the strict life of a Wesleyan Methodist under the guidance of his disciplinarian father. We will discover later that Moses became a non-ordained Methodist preacher, but whilst he was studying the Bible and considering its many teachings, he was also acquiring the means by which he could educate himself in astronomy and mathematics, the latter being an essential requirement for the study of the former. The fact that he became an engineer and a grinder of lenses in furtherance of his astronomical pursuits raises even more questions than there are answers. The fact that he succeeded in his many areas of interest is borne out in the results, and it is this success that we need to bear in mind as we delve further into his extraordinary life.

Regardless of the fact that Moses worked only as a handloom weaver as a child, a foundry worker in his youth and early twenties, and, after an undisclosed injury in the foundry, as a landscape gardener, there was nothing to suggest that he earned more than could be expected from such livings, but he was able to assemble a library of the necessary textbooks to educate himself to the extraordinary levels we will later discover. There is possibly little doubt that the assembly of the library was a dual effort with his elder brother, but even so it was a remarkable feat and quite unheard of for people in their station in life. There is a certainty to the fact that they, and particularly Moses, did achieve it against the odds, and furthermore they made full use of them, as later exploits gave testimony.

Before moving on, I would like to share with you a thought I have regarding the acquisition of the books that I earlier referred to. There is nothing recorded about his work as a landscape gardener, and yet I find it an unusual way of describing himself if he was simply a gardener. We were, for instance, many years away from Corporations providing landscaped pleasure grounds, so one can only presume that the work was in private grounds. We will see how Moses developed a talent for making contact with the more important people in the town, and he is noted as having said that 'In the early stages of his career, he found kind patrons amongst the rising manufacturers of Preston.' For instance, I have reason to believe that his foundry work was at one of Horrockses mills, so after he was injured, did he work at the homes of any members of that family? Did he find work at the homes of the professional people in the town? He later became a friend of Thomas Norris at Howick Hall, a man who also had an interest in astronomy, and perhaps had possession of the expensive written material that Moses sought. There is not a shred of evidence to support these ideas, other than Moses' single recorded statement regarding patronage, but I feel that it does have some credibility. He certainly seemed to have a special relationship with George Horrocks, the Turton-born member of the cotton empire family, who quite feasibly took Moses under his financial wing, for he, George, as we will see later, had a close interest in the heavens.

Turning to his involvement with Methodism, little is known prior to 1800, with the exception of him becoming involved, presumably through his father, at the chapel that was close to the junction of Starch House Square and Black Horse Yard, and sited behind what was later to become the Starch House Motors depot. It was close to this spot that he attended a sermon given by John Wesley in 1790, and a commemorative blue-plaque on the wall of a night-club and opposite the Market Inn, Preston, records Wesley's sermon there on the 15th April, although neither the inn nor the club were there at the time. It was at this chapel that Moses began to earn a reputation as a preacher, with a voice that was later described as 'deep bass and resonant.' It was added that his accent 'contained a considerable patois of the district in which he was born', but the pragmatic Moses was later to declare that he could never be convinced of a good enough reason to make the effort to change it. Whether his Lancashire accent favoured the totally distinct Bolton or Preston versions or an element of both isn't known, but every credit to him for his resistance to modify it. Perhaps he had a different style of delivery when lecturing rather than preaching. In those circumstances, I have seen him described as having a homely, pleasant, conversational way of imparting information, and because he avoided as far as possible, the use of technical terms, his message became more impressed on the minds of his audiences.

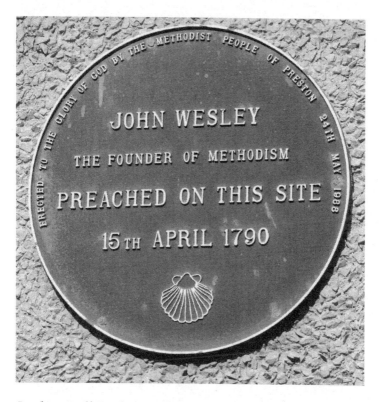

In the small Back Lane Chapel, John Wesley preached on the occasion of his last visit to Lancashire. The scene on that occasion is described as having been most impressive. The veteran preacher, who was then fast approaching his 88th year, having ascended the pulpit, announced his text, 'The Spirit and the Bride say come.' Then, it is said he lifted up his hand to his forehead and paused for some time before he could proceed. His memory had failed him.

It may also be useful to imagine what Preston was like when the Holden family first arrived in the town. The population was a mere 6,000, rising to around 10,000 by the year 1800, as the mills began to grow and spread. The standard of life endured by the population was for the most part unacceptable, and the immorality appalling. Cock-fighting, dog-racing, bull-baiting, pigeon-flying, drinking, and impiety abounded, wrote W. Pilkington in his book, *'Makers of Wesleyan Methodism' (1890)*, who then went on to say that Christianity was certainly at a low point, and furthermore there didn't appear to be anyone who was prepared to address the problems or to influence the spiritual well-being of the population. Pilkington described a large part of that population as 'ignorant, profane, filthy, and disorderly in the extreme'. It was around this time that the town saw the advent of a group of pious men, not necessarily men who were highly educated, but men who "had clear heads, warm hearts, and a burning passion to rescue the perishing and save the dying." These were the men who "blew the gospel trumpet, and aroused an excitement which shook the town and its neighbourhood. They were the Wesleyan Methodists, and Moses Holden was about to join their active ranks.

There is a slight irony in recording that one of the earliest centres for Methodism was a public house on Church Street, the Old Dog Inn. From the mid-1700s the landlady was Mrs Walmsley, the widow of John Walmsley, and herself a Methodist. She made a room available on the upper floor of her house for her friend Martha Thompson and others of her persuasion to meet and pray. John Wesley himself was to visit there on a number of occasions, leaving his horse in the extensive stabling facilities whilst staying in the town, and a blue-plaque records Martha Thompson's efforts here. Mrs. Walmsley's son went on to be ordained as a Methodist preacher.

Martha Thompson Blue Plaque

One of the attendees at the Old Dog Inn meetings was the son of Michael Emmett, landlord of the Ram's Head Inn in Gin Bow Entry. He too was called Michael, and he also became an ordained minister of the church. Michael Junior married the daughter of Roger Crane, a leader of the Methodist Society in Preston, and it was in 1795 that Moses was introduced to Crane. He became influential in Moses' life in the following several years, a period when Moses was to become a popular lay preacher. By this time, also, his father, Thomas was a class leader at the Back Lane chapel.

We will also discover how Moses played a huge part, not only in the founding of the Institute for the Diffusion of Knowledge, but his efforts with the Temperance Society, preaching to raise funds for all manner of church-related projects, and his involvement with Bible Societies and Samaritan groups.

However, the main focus of Moses' life was in connection with astronomy, and we will discover that after he had thoroughly grasped his subject, he was able to undertake a tour of lectures, reputed to be 'of the principal northern towns and cities,' but one of the earliest of those was at the Theatre Royal, Nottingham, in 1817, and that can scarcely be classed as a northern town. So how far did he travel? I'm convinced there is still a huge amount to learn about the extent of them. In the previous year, 1816, he'd given his series of lectures in the February, at the Leeds Theatre. At first glance he seems to have crossed the Pennines for these events, but his son, William Archimedes was born in that year in Pontefract, so maybe they lived in Yorkshire for a while? There is no evidence to support that theory, and Pontefract is about seventeen miles from Leeds, so in those days, three hours distant or more.

To confuse matters about the distance and extent of Moses' travels, I am drawn to a sermon that was delivered by the Rev. Hugh MacNeile in Preston Parish Church. The sermon was titled

'The Astronomer and the Christian,' and Moses Holden was the subject used as the character in the story. During that sermon, the Rev. MacNeile said of Moses, 'While lecturing in the southern counties, he was frequently asked about Jeremiah Horrocks,' perhaps an indication that there is still much to learn.

To supplement his talks with the visual aids that would help to bring it all to life, he constructed an orrery, a clockwork instrument that demonstrates the movement of each of the planets, one with the other. They can be simple ones that include just the Sun, Moon and the Earth, or a more complicated one that includes all the planets. Moses chose the complicated one! He also constructed a magic lantern, or more probably, a number of magic lanterns, to use in conjunction with the orrery. In addition to this he constructed telescopes for both himself and others, and ground and polished lenses for use in those telescopes. As I mentioned earlier, researchers into the life of one of Preston's earliest photographers, Robert Pateson, tell me that it is almost certain that Moses taught him how to grind and polish lenses for his own use in photography. For a more detailed explanation of what an orrery is, please go to Appendix Seven.

Before moving on to the next stage of Moses' life, it may be appropriate to take a look at two stories that help to give us an idea of his character and personality.

His upbringing and religious convictions conspired to prevent him from ever watching a stage play, regardless of the pressure that was applied from time to time by his friends who he held in high regard. When his son was about thirty years old, he joined others in a production of Hamlet at the Theatre Royal, in aid of the fund for the purchase of Shakespeare's house at Stratford-upon-Avon. William took the part of the Ghost. Despite holding the works of Shakespeare in high regard, and though he didn't express total disapproval of the performance itself, Moses refused to go and watch it. In reply to one of his friends who had attempted to change his mind on the matter, he replied, 'In my younger days, I refused a five pound note and a box ticket offered to me if I would go and see Mrs. Siddons at Drury Lane; so I think I will not break my resolution at my time of life for what you amateurs can do.'

Despite that particular objection, there is plenty of anecdotal evidence to suggest that Moses had a keen sense of humour, and something that can be no better illustrated than in the story of an irreligious old woman he came across. He had for some time tried to make an impression on her to help her improve her ways and her life, but made little headway. One evening, seeing the old lady was sitting alone in a darkened room at her home, he placed a glass slide in his 'magic lantern' on which was painted a human skeleton. The image was projected through the old lady's window into the room in such a way as it appeared as a small speck of light on the wall. Then, 'by the usual process he gradually increased it in size and distinctness until a life-size image of the skeleton, with all its ghastly proportions appearing before her astonished eyes.' Moses reported that the old woman was so terrified that she became an exemplary character from that time to the end of her days. The authenticity of this story cannot, of course, be proven, but I have a suspicion that it was the sort of thing he may do, in his attempt to return someone to the straight and narrow ... or even take them there for the first time. This type of projection of images was the process that was commonly referred to as 'phantasmagoria.'

Chapter Two

1801 – 1810 First Evidence of Preaching

The opening decade of the nineteenth century, saw the death of Moses' mother, Joyce, aged about sixty four years on the 27th January 1805, and his fellow astronomer brother, John, on the 23rd January 1810, at the regrettably young age of thirty-five years. He was described as an astronomer and mathematician, which would suggest that the two brothers were closely linked in their pursuit of knowledge. The journals, which I will be referring to in the next chapter and begun barely twelve months after his brother's death, contain no mention of him at all.

There is no doubt at all that Moses was a pious man, and that in the early days he remained loyal to his mentors, including his father, Roger Crane and other leaders in the Methodist Society of the 1790s, and the Reverend Thomas Jackson, the Superintendent of the Methodist Circuit, but as his life progressed I have formed the opinion that the name of the religion mattered less and less to him. His final thirty years were spent as a member of the Established Church, forming a close relationship with the Vicar of Christ Church, Preston, The Rev. Thomas Clark, although he frequently acquiesced to the wishes of his old friends, and preached charity sermons in various Dissenting chapels. Despite that, in the April of 1846, Moses was elected as a churchwarden of Christ Church

I have formed the impression that this was a decade of preparation for twenty three year old Moses, through to his 32nd birthday. It was a decade of doing his duty as a Methodist preacher, as well as one of following his passion as a practical astronomer. There is enough evidence from the very beginning of the next decade to confirm that Moses was already constructing telescopes both for his own and others' use. Indeed, I would suggest that this was possibly one of the ways that he was able to fund the purchase of the expensive books and other items he required to pursue his dreams. Unfortunately, there is very little documented evidence to prove things one way or the other.

It was also the first decade in which there is any record of Moses' long history as a correspondent with the press. The vast majority appeared in the 'Letters to the Editor' columns, so it is unlikely that they were a source of revenue. What they would be was a way in which he could raise the profile of his subject, with the intention of raising the interest and enthusiasm of his readers, and an advertisement to those that would be capable of buying a telescope from him. The local newspapers in Preston were the usual recipients of this correspondence, but other newspapers would use the material, crediting where it had first appeared. Eclipses of both the Sun and Moon were probably the most common and the most simply observed without a telescope, but the sighting and predictions of lines of travel of comets were also regular. Indeed it was one such comet that featured in the earliest letter I have seen, this being to the *Manchester Mercury* and dated the 10th November 1807. It also tells us that he is living temporarily back in Bolton at Ridgway Row in Bolton Moor. There were unconfirmed reports that the family had moved back to Bolton temporarily, and for a reason unknown, and I have reason to believe that Moses sustained the injury referred to in Chapter One at a foundry in Bolton.

The correspondence related to a comet that was capable of being seen for an extended period, certainly in excess of five weeks, and Moses gave drawings of its position in relation to other stars on six occasions he observed it between October 19th and 27th that year. He went on to say that 'some astronomers have supposed the comets to be made for no other end than for fuel for the Sun; if so, why do they go round the Sun and forsake it? 'But I can see no reason to believe the Sun to be a body of fire.' He based this observation on the fact that Mont Blanc and other high peaks retained their snow-cover throughout the summer months. Things have moved on somewhat from those early days!

Moses' work as a Methodist missionary began around the middle of the decade, probably in 1804 or 1805, although local historian and newspaper proprietor, Anthony Hewitson, writing some fifty years later in his church and chapel series of articles under his pen-name Atticus, gives 1810 as the date. He does give a little flexibility to that statement in his article, and after visiting the Quaker Movement premises in Lower Lane he went on to say, "From this place we proceeded to the Wesleyan Chapel, situated on the higher side of the village, close to the high road leading to Kirkham. It has a rather genteel exterior; and, as a small village chapel, looks prim and orderly in front. Wesleyan Methodism was established in Freckleton in 1810, and in those days 'Old Moses Holden,' of Preston, used to visit the place and astonish the natives with his strong rough-cut eloquence. Moses was one of the pioneers of Wesleyanism in Freckleton, along with Messrs P. Taylor, G. Richardson, and J. Grayson. The erection of a chapel succeeded the labours of Moses and his colleagues, and about 1840 the present building, which will accommodate 200, was erected on the same site."

W. Pilkington, in his 1890 book, 'Makers of Wesleyan Methodism in Preston,' recorded that 'somewhere in the first ten years of the 1800s, Moses Holden, who was then a popular

Preston Methodist local preacher, paid a visit to this ancient village, (Freckleton), and preached from the steps which led up to a room over the old smithy on Smithy Green.'

It is possible that Freckleton was ripe for a change from the domination of the Quakers; for Atticus recorded a conversation he had with a villager, who said that, "The meetings are too quiet. Freckleton fooak likes a good noise, and a lot of singing and th'Quakers don't go in for music, so the business is rather flat."

Returning to the suggestion that his first visit may have been in 1808, it is interesting to note that, in the report of the Lancashire Congregational Union of that year, the following description appeared; 'The Fylde is morally the most dark and miserable part of the county. Long ago did its wretchedness excite the sympathy of certain Christians who knew its situation, and a few attempts were made from time to time to diffuse throughout it the light of the Gospel ... Till of late little more could be done than to pity and to pray for this region of darkness.' We will discover more about Moses' exploits when he began an 18 month missionary tour, basing himself in Poulton, probably the epicentre of the Fylde and Over Wyre regions, this area was still, according to the records, 'an undesirable place for the Dissenter'.

I can offer a little to the debate about when Moses began to visit parts of the Fylde from what he wrote in his journals whilst on his tour of the Fylde in 1811 and 1812. After giving a sermon in Kirkham in early 1811, he received a request to visit the village of Bryning from a resident of that place. He wrote, " ... then set off for Bryning according to my promise. I got there, and met with the same homely, friendly and good reception I did there six or seven years since." He followed this by adding, "I desired them to inform Warton singers as they had been fond of me, and I knew they would come if anything of their respect remained. I told them to tell them that we would sing the Psalms in the

Prayer Book, the same as sung at Warton Chapel. They came and brought their Prayer Books with them, set off their tunes, and seemed uncommonly pleased. We had a good number and I felt it easy to speak, with a degree of power and liberty."

Bryning-with-Warton is now a single parish, but in those days was, along with Freckleton, close neighbours one with the other. There is so much more that remains undiscovered, for in my view it is unlikely that any group would 'become fond' of anybody after a single visit.

Chapter Three
1811 – 1820 **The Missionary Years**

This decade had hardly dawned when Moses embarked on an Evangelical missionary tour that was to last for eighteen months. As we have already discovered, Moses had already earned a reputation as an orator and spreader of the gospel, and this fact had not escaped the attention of the Rev. Thomas Jackson, the then Superintendent of the Preston Methodist Circuit. He had been attempting to encourage Moses to take up the challenge for many months, and towards the end of 1810 Jackson's will finally prevailed. Moses remarked that, "From the time of the (Methodist) Conference of 1810, he (Rev. Jackson) never let me rest until I consented to go and try and open up the Fylde country."

And so it was, that on the afternoon of Saturday January 19th 1811, and having received the offer of a horse, Moses set out for Poulton, the place that was to become his base for the foreseeable future. He rode most of the way but was overtaken by darkness, but was met en-route by '... a generous young man that was my guide. Got to Poulton at 7pm.' The only clue as to why he chose Poulton as his base was because 'it was the only place where Methodism had made any headway, and that there were in that place, ten members.'

So began a series of three small journals that Moses kept while on his travels, journals in which he recorded the discomforts and personal privations of tramping here and there across remote country districts, but he never gave any clue as to where in Poulton he was based; but he did talk of 'being entertained at the home of John and Betty Tomlinson', sometimes spelt Thomason, with John driving Moses to various villages in his cart to preach. He did have a static base, for he later talked about having some of his books and other possessions delivered there by carrier. He also had a cousin on his father's side of the family, Thomas Holden, living in Poulton, for on his third day he wrote, 'I got breakfast at Thos. Holden's my cousin, and from there I went to Marton', but there was no record of him staying with him. In fact, towards the end of February 1811 he wrote, "I called at Thos. Holden's, and when I talked to him to attend (a service in Poulton) he seemed to take no notice." Nothing is known about how his cousin Thomas was employed, but later in the year he wrote "I read in Thomas's library till noon or after," which makes it sound as though he was leading a reasonably successful life. We will see later how he organised a fortnightly cycle, visiting small settlements and villages as he preached, making amendments to it as required. It has to be remembered that communication would be difficult, with his first visit to anywhere being unannounced, but accompanied by somebody who would be familiar to them. Thereafter, it could be relied on that Moses would be back to preach on the same day either every week or every two weeks. In the meantime, those who had joined his group would be expected to enrol others before his next visit, and so each group would continue to grow and hopefully prosper. He noted in his journal that on one occasion prior to visiting Kirkham he had sent a note to the Bellman, or Town Crier, to announce his visit, but he 'had gone out of town'.

The Rev. Kirkman (1885) in his 'Memoir of Thomas Crouch Hincksman', a Lytham man and Methodist stalwart, wrote with regard to Moses Holden's period in the Fylde, 'The mission seems to have been prosecuted throughout the Fylde, from January 1811, to August in the same year', a statement which I find confusing. It is my belief that the journals which I rediscovered in the Harris Reference Library in Preston in 2004, were at one time in the possession of Hincksman, and they clearly extend to the 29th May 1812. The second of the three journals continues until the 7th October 1811, so the comment is likely to remain a mystery.

The arrival into Methodism for John and Betty Tomlinson is worth recounting, for in the first place only Betty had been converted. She was the first member of the Methodist Society in Poulton after she had had a conversation with a group of devout women who had caused her to turn her own thoughts to matters of religion. It was said that on one particular day, whilst praying in a cattle shed, she was converted. She made contact with two or three pious women in the neighbourhood, and met with them regularly for the purpose of praying together. John, a butcher by trade, was at this point totally opposed to Methodism, and went so far as to lock his wife out of the house when she went to pray. How long he allowed her to remain outside isn't known, but the action failed to cure her, and his suspicions became aroused. He began to think that she was up to no good, and so set out to discover what it was.

He managed to conceal himself in the house where the women met, and spent an entire evening watching the proceedings without their knowledge. However he was unprepared for what was to happen at the end of the meeting when they began their final prayers. The women prayed for him by name, asking God to change his heart, and make him a new creature in Christ Jesus. He began to cry for mercy, and emerged from his hiding place a penitent, and conscience-stricken man, and joined his wife in her faithfulness and devotion.

When Moses arrived in Poulton in January there were about ten members in their Methodist Society, but they needed a leader. So Moses assumed that role with a class that included John and Betty (so it was after John's conversion), and four other people. Peaceful worship wasn't allowed to continue for long, however, in the small room they used as a chapel. A mob smashed all the windows with stones; they broke into the chapel and dragged the pulpit out into the market place. The mob's behaviour extended to disgraceful and insulting verbal assaults on the Methodists, but the more they persecuted them, the more the class grew, both in numbers and strength. Very little, if any of this behaviour found its way into the pages of Moses' journals.

The dramatic description of John's conversion brings to mind what Methodists referred to as 'weeping seasons'. The passion and fervour with which preachers delivered their sermons often reduced congregations to floods of tears, tears of self-guilt and repentance. On February 24th 1811 he wrote, "I saw my weaknesses and wept over it," and two months later, on the 23rd May he wrote, "I met the class and a weeping season it was."

Another term that recurs in the journals is that of 'Love Feast'. It had featured in several forms of Methodism for many years, and was adopted by John Wesley after his visit to America in 1737. Originally he introduced them on a monthly basis, but that was soon reduced to quarterly. They involved a sharing of simple foods and drinks, often sweet buns and tea, but different groups had other ideas. However, the essence of the meetings is to heal any relationships that have been spoiled or tarnished, and to forgive others of the things that may have affected you adversely. It was an opportunity to renew and hopefully strengthen relationships, and bring people closer together. Each attendee had to be a ticket-holder, issued by the individual society for attendance by their members at church services.

The certain promise of an eternity in hell was never far from the Methodist preacher's lips, and Moses' was no exception to that. In March 1811 he wrote, "I said oh yes there is a difference, for if a good man be killed in a marl pit, he would leave his toil and hard labour for a heaven of rest in a minute, but the wicked man would go from the marl pit to be tossed in the fiery waves of hell", and in July the same year he recorded that "I said I was afraid if she did not get her heart weighed with the grace of God it would be so light it would bounce into hell."

Marl pits appeared all over the Fylde countryside, and indeed in many other parts of the country. The spoil obtained from them consisted of varying proportions of clay and carbonate of lime, which was used as a fertilizer. The resultant holes filled naturally with water to become the familiar ponds that have been a part of our landscape for the past two hundred years.

Moses enjoyed attending the sermons given by ministers of other religions, almost always being critical of what he saw and heard. On May 19th 1811 he wrote, "This forenoon I went to hear a young man at the Calvinist Chapel (in Poulton). He had a tuft of hair on his topping that would have done credit to Tommy Tillground's bull. He had a beautiful countenance, and a nice looking young man could read with a nice grace – he had notes with him. Began with such pomp and style, and would bring proofs out of scripture, looked in his notes and told where it was but could not find it, looked back to his notes two or three times till his countenance fell – and such a preach I never heard – but wonders never cease."

Just one month earlier he wrote regarding his view of the Calvinist doctrine, "I hesitated not to say that it was a doctrine of devils and came from Hell, and that those that were propagating these things were doing the Devil's jobs and running the Devil's errands – I see the Calvinists come for no good."

At the start of the 19th century there was a great confusion of sects, and throughout the journals, in addition to the Calvinists, he mentions the Anti-baptists, Independents and

Swedenborgians, and comments on how they held doctrines that hindered the advance of Methodism and how he had attempted to combat their teachings.

It may be useful if, at this stage, I familiarize my readers with the fortnightly routine to which I have referred. It has to be assumed, unless he recorded otherwise, that all his travelling would be on foot. There were occasions when he travelled on horseback, and less frequently by shandray, a one-horse, sprung carriage. On these rare occasions he would preach from the shandray itself. From Poulton he would walk the short distance to Thornton Marsh where, certainly in the earliest days meetings were held in a cottage next door to the mill itself, and the home of Thomas and Nancy Greenwood, believed to be the parents of Betty Tomlinson who we have already met. As an example of the harassment meted out to the dissenters, small boys climbed onto the thatched roof of the cottage, and inserted a live goose into the top of the chimney pot. The tumbling, and presumably furiously flapping goose, ultimately appeared on the hearth of the cottage, looking extremely unlike a goose that was capable of laying a golden egg, covered with, and surrounded by, a small mountain of soot! The meeting was delayed to allow the smoke to clear. Such were the trials that had to be borne, and whether the goose had been stunned isn't known, but it was said that it settled down amongst them, squatting peacefully at their feet, until the meeting was over.

An indication of the standard of living for most of the country-folk of the Fylde can be drawn from the comment made whilst visiting one such place. Moses had been accompanied by an Edward Threlfall to a meeting in a cottage, and after rubbing his feet hard at the door to clean them, the occupier tried to put him at his ease by saying, "You may come in without so much ado, we are none so clean". Edward replied, "I know that you are dirty folks, and thought you needed no more!"

Of all his classes, Thornton Mill was arguably his most successful, for in his "Memoirs of Thomas Crouch Hincksman" the Rev. William Kirkman wrote, "Holden's preaching and influence must have had an effect on the members at Thornton, for by their industry and the goodwill of friends they were able to build themselves a chapel within eighteen months of his leaving."

From Thornton Marsh he would make his way to the River Wyre at Shard, still bridgeless at this point of course, and so he would seek the services of a boatman to row him to the Over Wyre or northern side of the river. It is of interest to learn that he recorded in his diary that 'he sought the help of the boatman when the tide was in,' presumably inferring that it was possible to cross on foot when the tide was at its lowest? In reality, I think low tide crossings would be effected on horse-back. During his first month in the Fylde he related a story where, 'I went to Wyre-side ... the tide was at its height and extended from bank to bank. I waited long for the boatman to come. The water was rough and the waves ran high, but the wind soon blew us over.' He always travelled on his own, carrying a small parcel of necessaries, which included of course, his Bible and his diary, or journal. Meals and a roof over his head for the night would be found with one of his flock, and from what can be established, it seems that he generally stayed with the same people on each visit.

I also get the impression that some individuals may have had a financial or other inducement to encourage others to attend classes and services, for on the 19th February 1811, Moses wrote, "I got Bartle to give his bond to Mr. Cookson, carpenter, to see him paid conjointly with James Parkinson for filling up Chapel the course of this year."

The first village in Over Wyre will have been Hambleton, but he makes little comment about it other than the time that he called at the Chapel Yard. "Saw a nice bush of thyme set on a grave. Felt solemnized by treading the turf of the silent dead. Sought for Miss Blundel's grave. Saw no inscription." A little further on towards Pilling we reach Stalmine, or, as he writes it in his journal, Stomen. This was the written form of the way the word was pronounced, and I understand that there are still elderly residents in 2014 who continue to say it that way. Similarly, Rawcliffe was pronounced 'Raycla' and Poulton was pronounced, 'Pootun', so one has to keep an open mind with regards not only to all the place-names mentioned, but also those of the people with whom he came into contact.

During the second year of his missionary Moses visited Stalmine on Shrove Tuesday, and made a note that "It rained, and when I got there through the wetness, or it being Shrove Tuesday, I don't know which or both, only two or three came besides what went with me. We had a prayer meeting." What was going through Moses' mind when he queried whether Shrove Tuesday could be the reason for a low turn-out isn't clear, but 'shrove' is derived from the word 'shrive' meaning confess, so whether, prior to Lent the country folk around Stalmine weren't prepared to come and air their shortcomings will never be known.

Travelling closer still to Pilling we find Preesall where, on his visit on Wednesday 13th March 1811, he recorded the wonderfully romantic story of walking to the top of Preesall Mill Hill where he looked over in the direction of Sunderland Point and Morecambe. Although the old windmill still exists on the hill, the same view cannot be had today because of housing and other developments. He then told how he saw 'a new vessel going out of Lancaster, a great size, above 500 tons burden, touching forty two yards long; and there were fourteen or fifteen other small ships, brigs and sloops all in view; a very beautiful sight.' I have attempted to replicate this view by finding a high point on the northern side of Preesall, and the distance to the mouth of the River Lune is such that I find it incredible that he was able to make the judgements about the vessel's size without the aid of a telescope; perhaps he'd already made a telescope, but I think he would have made at least passing reference to it in the journal.

Out of interest I enquired at the Maritime Museum in Lancaster, and discovered that there were only two vessels launched in Lancaster in 1811, both by the shipbuilders Brockbanks, and both owned by Messrs. Bolton and Littledale of Liverpool. They were the 495 ton 'William Danson', and the 411 ton 'Elizabeth'.

Speaking of the sea and of ships, Parrox Hall in Preesall was the home of Captain Elleston, a man who Moses went to see in the April of 1811. He described Elletson as a Guinea Captain, which was a euphemism for a slave trader. Indeed, Elletson related a story to him about a black man who approached him to sell him some gold, but Elletson was suspicious and had it examined. It transpired that it had a high content of lead, so those about him were ordered to seize him and lay him in irons. Having so done, a council was formed and his relatives came to see the accused. Moses asked him if he had paid, and he replied, "No, but the decision of the council was that the prisoner must find one prime slave, man or woman, one bullock, one fat sheep, and two ounces of gold." Moses asked him if he took it, and he said that he had.

It is of interest to note that Captain Elletson's brother was married to the sister of George Holden, the compiler of the Liverpool Tide Table in succession to his father and grand-father. It is to my knowledge that some members of Moses Holden's family have tried to establish a link between the two Holden dynasties, and whilst not dismissing the possibility, I'm afraid any link may be impracticably distant to be realised.

Pilling Lane End, and more rarely Pilling village itself were the next areas visited, where Moses occasionally recorded foraging at Sandside on the cockle-scars, or cockle-beds among the low rocks in that area, for a veritable feast of fresh shellfish. Morecambe Bay, of which the Pilling area is a southerly constituent, is still one of the pre-eminent cockling areas in the country. When preaching in Lytham he spoke of visiting the mussel-scars for the same purpose.

On his way back towards the River Wyre he always visited Rawcliffe and Out Rawcliffe, and it was at this place that John Taylor noted in his "Apostles of Fylde Methodism", that Moses' persecutors "let loose a vicious bull-dog to worry him in the street; but Moses stood perfectly still in the middle of the road whistling the 'Old Hundredth', and staring the brute out of countenance. It came close to the good man's feet, and then slunk away without touching him.' The Old Hundredth is probably better known as the tune to which a congregation might sing the hymn, "All People that on Earth do Dwell."

Although no mention of it is made in the journals, Moses probably re-crossed the River Wyre at Cartford Toll Bridge, by which the ancient Cartford Inn has stood since the 1600s. Eccleston, probably a combination of Little Eccleston and Great Eccleston, only remained on his circuit for a short period of time for in a revision of his timetable he left it out completely. One can only assume that he had been met with a negative reception, for this would be totally out of character for a man who would never say die. From that time onwards he would not have the need to cross the river, and may well have visited the tiny hamlets and farms to the north of the river, such as Moss Side. Places called Moss Side are encountered several times in the Fylde and Over Wyre area, but this one was a small group of dwellings just to the north of St. Michael's which he refers to many times. He had encountered them on his first visit to St. Michael's, a visit for which he recorded that he was met with a cool reception. He had made up his mind to leave the very small gathering, but an old woman had said, "Go not, but go with me, for I am sure that if some people on the moss knew, you must not go without preaching." With that he was taken to visit the Mallys of Wescot or Westwood. "They were ready to jump for joy" he wrote, and they filled the house for him to preach, and promised to do the same if he were to call each fortnight in the future.

Moses kept his word, and just a fortnight later he again visited the same people at Moss Side and Wescot. He wrote, "I was received with the homely welcome of country people, but the night was unfavourable, it snowed and blew a storm, and few attended." It appears that there were people present who held their suspicions. He continued, "After I had preached an alarming sermon, I joined four or five in class, but I think the people seemed terrified at the mention of a class. I don't see any other cause than they think that there will be some money going." He went on to explain that he would, as a Christian, only expect them to pay within their ability to do so, finishing his entry for that day with the simply sentence, "The widow and her two mites", and one can only imagine the desperate situation of these three individuals.

After visiting St. Michael's he would make his way to Garstang by way of Churchtown, where, on one occasion he called at St. Helen's Church of England church. It was to Moses' knowledge that there had been some extensive lowering and rebuilding of part of the roof of the church, and he felt the need to go and inspect the work. He recorded that while he was there he climbed the internal stone spiral staircase to the flat roof of the tower, and, in typical Moses' fashion he recorded that "he surveyed the whole of the Fylde area" from his prominent perch. I attempted to replicate that exercise, but failed to take into consideration the growth made by a number of beech trees and their attendant foliage in the intervening years. I saw little or nothing!

St. Helen's Church, Churchtown.

The stonework that supported the original roof of the church can be seen on the side of the tower on St. Helen's church, Churchtown, near Garstang.

Early in his mission it was implied that there may be an opening for him to start a class in Churchtown, but after meeting Mr. Lilley, the Superintendent of the Lancaster Circuit in Eccleston, he was told that the Lancaster preachers would not give it up. With geographical arrangements of this nature, there is always likely to be some overlap. However, as early as the 28th March he wrote, "Set off for Churchtown. John Mally went with me. I preached to very nigh a large house full." He had intimated that he thought he would be well received in Churchtown when he met Mr. Lilley in Eccleston, so perhaps he'd set up a rival class. It would also answer why Moses wasn't accorded a warmer welcome.

There is nothing recorded about any activities in Garstang, although we will see later that he did visit the village. There was an occasion where he recorded having visited Scorton and another when he had walked across the fields to Oakenclough to preach. On his return journey he recorded that he came past Greenhalgh Castle on the outskirts of the village, referring to it in the way it is pronounced, "Greenah Castle, the place knocked down by Oliver Cromwell."

Another place to which he made reference was the irregularly shaped building to the east of the Roe Buck Inn at Bilsborough and Bilsborough Parish Church. It was known as the Pot House, and Moses made a number of attempts to have it converted for use as a Methodist Chapel, but his ideas had not come to fruition before the end of his stay.

Pot House, Bilsborrow

It has been said that the Pot House was converted to a chapel in 1811, but this is erroneous. It would be towards the end of 1812 or even a little later, and well after Moses Holden's mission had come to an end.

From either here or other parts of Garstang his journey was undertaken in a slightly different way. He made full use of what was a relatively new Preston to Lancaster canal, and from here he would take the barge boat to Salwick, from where he would either walk or ride a horse to Freckleton, the one place he'd preached before embarking on his eighteen month mission. Probably his final visit to Freckleton during the period of this missionary was a few days after he had learned that his father was dying. He had arrived in Garstang to find a letter waiting for him. It carried the news that if he wished to see his father alive, he must go immediately to Preston. He went, and found him "weak, but sensible and happy." He stayed with him until the Sunday, when he left in the knowledge that his father was comfortable and confident about what was the inevitable outcome that awaited him, and that he would be unlikely to see him again 'in the land of the living'

It seems that duty called, and he recorded travelling from Preston to Kirkham with Sergeant Reynolds. I have no idea who this individual may have been, and I don't even know whether 'Sergeant' was a Christian name or a rank. Either is a possibility. Later the same day he recorded going to Freckleton, writing, "In the afternoon, never did I feel so sad or heavy, although I had felt happy as I came along the lane; but nature seemed to mourn. I went to Freckleton. Mr. Wright had got the promise of a barn in which to preach, but we could not get in, nor could we get a key. I stood on a chair in the open air, and very cold. My flesh turned blue as if stained with logwood." I find that Logwood is a tree that is native to Central America, and had long been used in dyeing. Its scientific name of *Haematoxylum* would seem to suggest that the dye would be blood-coloured, and one of its common names is 'Bloodwood', but if grown on alkaline soils rather than acidic, the dye produced becomes blue.

On Thursday 20th February, 1812, Moses recorded that he 'went to Garstang after dinner'. He had been in Pilling the previous day, and it would have been unusual for him to travel directly between those two places. When he arrived in Garstang, he wrote, 'Mr. Ogle came and gave me a call. He said my father was much as I had left him, but soon after I had a letter that told me that my father was dead. I set off for Preston. Though nigh night, I got there by nine at night. I strew the lane with my tears, but in the midst of this I felt a firm confidence that he was landed in rest. He died on Wednesday 19th February, 1812, 6.02pm. I felt a little weighed at the thought of having lost my last real friend and counsellor. I could but say well, he is happy, forever happy."

Moses' father was interred just a few days later, on Saturday 22nd February, but it wasn't until Sunday 22nd March that he wrote, "I preached my father's funeral sermon at Garstang (on that date) to a great concourse of people." There seems to be no reason why it should have taken place in Garstang, and no clues as to why were offered in the journal.

A neighbouring area to Freckleton was Bryning, the place written about at the end of chapter two. Although I mentioned that the people here became fond of Moses, it seems that all were not so inclined. It is similar to the unrecorded story told by John Taylor in his 'Apostles of Fylde Methodism' in Rawcliffe about dogs assailing him, but on this occasion Moses recorded it in his journal. It occurred on Tuesday 2nd April 1811, near to Bryning, and he wrote, "I then set forward, and when I got half a mile by the farmhouse, there came two dogs. A large Mastiff and a surly cur. I thought they were set on me by somebody. They came like hunting dogs. When they began the hideous howl, I just could hear them, but in a minute or two they were with me, their bristles stood up from their head to their tail. I stood still, and they stood about three yards off me. They might stand roaring about three minutes. I then whistled a tune. A nice air. They gave over barking, dropped their bristles, and I went on, and they went off. I was not within call or sight of anyone."

The next time that Moses passed this way he called at Carr-side, the farmhouse concerned, and told the occupant that if she allowed her dogs to come after him again, she would be fined.

Kirkham would be the place visited after Freckleton, but he seems to have been only marginally successful in this small town that John Porter described in his 1878 'History of the Fylde of Lancashire', "....probably the earliest inhabited locality in the Fylde district." It is certainly pre-Roman, and its history could stretch back 10,000 years.

After one of his visits to Kirkham, Moses wrote of meeting a woman that he had known from long ago. He said, "Jane Shakeshaft came by and they asked her to come in. She had been a Methodist about seven years, and a lively one. She, I believe, enjoyed a real religion then. She came in and I said, 'How do you do Jane?' She seemed thunder struck. I said to her, 'You once were alive to God. How is it now?' She said, 'I am very wicked. I never read or pray. I have got a deal of wicked companions. I go to the ale house and swear and dance'. I talked to her, and she seemed petrified to the floor. I begged her to come and hear a sermon next time I came. She said she would. She said she felt miserable, and I told her it was a sign that God had not given her up, but that her companions were her chain. I told her to talk to them about their souls and they will love you." Although Moses doesn't say that *he* knew her seven years previously, it is possible that he was actively preaching in the area as early as 1804.

On one of his visits to Kirkham he wrote of preaching to a group of around seventy people, including a group of boys aged fifteen or sixteen. He wrote, "There might be thirty or forty of them. They behaved well. Spoke to them till they cried. Apprentices to A. Berlow. They are from the Foundling Hospital, London." A. Berlow was a mill operator in Kirkham.

From Kirkham he would make his way to Lytham, although which route he took isn't known. He may well have varied the route, but nothing was recorded to indicate that. One obvious route would be through the village of Wrea Green, mentioned in the Domesday Book, but not in Moses' journals.

However, whatever route he chose to get to Lytham, it is worth remembering that in the early years of the 19th century, Lytham was little more than a few fisherman's and shrimper's thatched cottages. Virtually all the land and buildings were controlled by the staunchly Catholic Clifton family of Lytham Hall. When Moses arrived in Lytham for the first time, he wrote in his journal, "I found a house licensed for preaching". It is understood that a Mr. Lyon had got it licensed, and preached in it for a long time before; but it had been given up. However, it opened again, and had good congregations, but a deal of persecution. The clergymen of the parish came to the house and kicked up a great stir, demanding the licence. They made several threats to the gathered assembly, and eventually went to see Squire Clifton, asking him to put a stop to this Methodism, and get Holden sent out of Lytham. However, the Squire is said to have replied "I shall do no such thing. Let them alone or I may put Holden in your place." This story was preserved only orally, but he was untroubled following the unlikely intervention, but unfortunately, Moses later wrote regarding his efforts there, "I never could prevail on any at Lytham. They received me kindly, and heard me gladly, but that was all."

On another occasion he recorded arriving in Lytham around noon, and being informed that "Lyster, the Church Minister, had come raging the day before, demanded to know my name and demanded to see the licence (for preaching). He said he would write to the Bishop. I see the Divel is stirring. May God choose me and give me courage. We had a house full; they behaved well." The Rev. Robert Lister was the minister at Lytham for upwards of thirty years, starting in about 1805.

During the mid-1840s a preacher from Manchester, the Rev. Thomas Crouch Hincksman, brought his family to stay in Lytham for a time, and formed the impression that Methodism ought to be represented there. He had, at one time, been an active member of the Wesleyan community in Preston, so he approached the Preston Methodist Circuit with the promise of a subscription from himself, and further subscriptions from 'influential men' in Manchester. At the same time, another Lytham resident became active in the raising of funds for the same purpose, that being the erection of a chapel. The Circuit applied for land for the purpose, and it was kindly granted by Colonel Clifton. A contract was drawn up for a chapel that would cost around £600, and the Colonel determined that it should be erected on exactly the spot that was once Mercer's simple thatched cottage where the Gospel had been preached thirty-five years earlier. In the intervening years the land on which the cottage stood had been cultivated, with it being described at the point it was transferred as a cabbage patch. It was opened in 1847. Furthermore, Colonel Clifton provided the land for a minister's house, which like the chapel was on a 999 year lease.

Before moving on from Lytham it's worth saying that although Lytham was something of a failure overall, the majority of one family, the Mercers, had openly declared their allegiance to the Methodists. Their name appears many times in the journals, but sometimes written as I suspect it was pronounced, 'Massor'. It was in James Mercer's low thatched cottage in Bath Street that Moses held his meetings, and was the one referred to as the place originally licensed by Mr. Lyon. Mercer was a fisherman, probably a shrimper, by occupation, but although it was still referred to as 'James Mercer's house', it is likely that he was by this time dead, but his family still occupied it.

On the evening of March 5th 1811, Moses recorded being at Mercer's cottage where he met two old men, one above eighty years and the other above eighty five years, who had been neighbours from childhood. He said that they could remember a lot of ancient things, including the sighting of a Comet that had been visible for a long time. They said its tail was as long, or longer, than a church steeple, and one of them said that its nucleus was the size of Venus seen with the naked eye. To compare something to the size of a church steeple seems to me to be an unsatisfactory way to describe something. A church steeple viewed from how far away? However, we probably know what he meant.

Later the same evening one of Mercer's sons, who was almost blind, insisted that he be allowed to show Moses something that his father had written and left to him. They were a collection of hymns that had been called, 'The Fishermen's Hymns', and were a collection of twenty-four sacred songs on Gospel themes. In the 'Address to the Reader' at the beginning of the work, James Mercer had written, "The hardships and difficulties which I have experienced in a low occupation and laborious life for the support of my family, called me to the frequent reading of the Holy Scriptures; from this fountain of true comfort under every calamity, my scanty meals and hard lodging became as delicious feasts and a bed of down."

On his journeys from Lytham to Marton, Moses made several references to Peel, probably better known to more modern generations by the neighbouring Peel Corner, a well-known landmark on pre-motorway journeys to the illuminations and other seaside pleasures. We will also read more about his visits to Marton, which at the time would include Peel, remembering that they were all part of the Parish of Poulton-le-Fylde long before the birth of Blackpool.

The final visit on his fortnightly call cycle was to Bispham. The Gynn Square area of Bispham was virtually all there was of Blackpool. In 1801 Blackpool had a population of just 500, just five per cent of the population of Preston at the same time. Despite its size, there were many interesting visits to Bispham to record in his journals, and on one of which he recorded that he had been opposed by the incumbent, the Rev. William Elston, a dour parson who had been born at Mythop to a well-known local family, and who had served for forty years in Bispham. He wrote, "One day when I had to preach there some of the people begged I would not go … for Mr. Elston said he would have me put in prison. I said that would be an honour. I went and preached without any disturbance."

On another visit to Bispham, Sunday 16th June 1811, Moses wrote "We heard as we were going that the people had made it up to stone us. It certainly was so, but several men and women went with me. I got a chair the next door (sic) to the place that let us have it the last time. Several came near, and some stood at a distance. Yes, a deal of people was in the town waiting. I began, and gave out a hymn, and shouted to them at a distance, and said 'You had better come nearer' (I thought it was them that had come to disturb). But I said, 'You might be afraid of an uproar, but you don't need, for there dares not one to disturb us, for Government protects us'. They behaved very well. I rode back with Mrs. Campbell and Betty Tomeson." Note yet another spelling of 'Tomlinson'.

At another time, James Morrow, the Congregational minister of Poulton, and a man who Moses had conversed with regarding theological matters on several occasions, sent word that Moses had better not go to Bispham for, "… he was well-assured that the clergyman had engaged several men to kill me, and they were to have ale and rum mixed, to fit them for their work. Many came to persuade me not to go … for the clergyman had threatened everyone who either lent me a chair or allowed me to stand on their horse block."

Typically shunning the well-intended advice, Moses went to Bispham in a shandray, a form of pony and trap, presumably driven by John Tomlinson, and he used the back of the trap as a pulpit. True to the warning a number of men appeared whilst he was preaching and a number of those present became alarmed, but Moses, as on a previous occasion, invited the men to come closer and listen to what he had to say and he told the crowd that they need not worry about any disturbance since he would be protected by the law of the land and by God. Remarkably, after hearing this, the hired-hands who were expected to cause a disturbance, were stopped in their tracks and stood quietly during the whole of the sermon.

Having looked at the framework of Moses' circuit, and noted one or two interesting events, let's go right back to January 20th 1811, and his first full day of what was to become quite an adventure. It was a Sunday, and like most Sundays he preached twice, although there were several when he would do so three times. In the morning he spoke at Thornton Marsh, a place where, certainly in the early days, he was sorely tried by those who disapproved. We have already learned of a goose being dropped down the chimney, but a rather less dramatic example came from a woman who tried to belittle him. He was well known for wearing a very good overcoat which possibly gave him a haughty or arrogant appearance, so they tried to make him wear their more humble garments to preach in, something he was quite happy to do. An old woman in a red cloak once said to him, "Will you have my cloak?" and he replied, "Aye, and your bonnet too, if you like." None of this kind of event found their way into his journal, and come down to us from others who were present at the time with their oral accounts being recorded later by John Taylor and others.

Just three days after arriving in Poulton, John Tomlinson, the character with whom Moses is thought to have resided, took him over the River Wyre on his horse, and took him as far as Preesall. Where possible, Moses made his first visits to places in the company of somebody else, perhaps to make the introductions. He arranged to stay the night with a Mr. Neal and his family, a man whose name similarly recurs throughout the journal, in various spelling forms, and occasionally with a 'Mc' preceding it. Probably bearing in mind the reputation with which the whole area had been disparagingly described, it is interesting to note that so early in his ministry he wrote of Neil's home, "I was astonished to find such a family in these parts. Oh, what heavenly simplicity dwelt among them. At night we knelt down. I prayed, then Mr. McNeil, his two sons and wife. I then concluded – but there were two little girls. McNeil said, 'Lasses! Are you asleep?' One of them began to pray. I was so affected by the power she prayed with, I could but weep. No sooner had she done but another began. That astonished me as much."

He continued the following morning by recording that "I was awoken by the family singing a hymn about half past five. When I got down, one had prayed. There was the father, three sons, and two wives." It is assumed that the two wives had different husbands, but much of the journal is in the form of notes rather than fully explained and grammatically correct.

Just a week after beginning his missionary work, Moses needed to finish off some outstanding work he had been doing for a man at Moon's Mill in Walton-le-dale. He had been waiting for some material being delivered from London in connection with a telescope he was constructing. He wrote, "I set about my work; I began to polish a large mirror or speculum 8 inches diameter, to a radius of 20 feet, 10 feet focal length, and finished it, but the metal was not mixed. Had yellow specks in it. Went to philosophic meeting." This was the 1810 – 1819 Preston Literary and Philosophical Society of which he was a member.

The following day he wrote, "Began to fit up a refractor, 9 feet long, 4 inches diameter," and the day after that, "I completed it as to fitting up, and had a look through at the Moon and Jupiter. I saw Jupiter's belts and satellites. This day I began the other two."

At the end of the five-day break from his mission, it would seem that he had completed the construction of two telescopes, started the construction of two others, and wrote, "I have been implored in getting things ready for another for myself. Sent my books to Poulton."

However, it wasn't long before he had another break. Perhaps he needed to construct and sell telescopes to provide the money for his missionary work, but on this occasion he was making equipment for himself. He was making new eye-pieces for an existing telescope, and at the same time re-modelling the telescope so that it could be broken down into two separate pieces. Furthermore, the stand on which the telescope stood was made so that it would dismantle. He also constructed a box that was capable of holding all this equipment, so that he would be able to move it from place to place with relative ease.

With those two breaks, Moses seemed to have completed any outstanding work at home, and although he did visit Preston a number of times afterwards, it was for things other than his astronomy interests. He spoke of visiting his only sister for dinner, and although he doesn't name her in the journals, I now know her to be Alice who was around four years his junior. She married a William Broadbelt, and at one point in her life with him (1841), they lived in Lune Street, Preston. On another occasion, on the 18th July 1811, Moses recorded that 'I set on to Preston this forenoon. I went to the Court, being serious for a licence to preach. I took an attorney with me; he spoke to the Clerk of Peace who rose and said, "Where is the Methodist Parson that wants swearing?" Three others came in, and it was all done as one; we gave him a half-crown each.'

Frustratingly, Moses written records are short on essential detail; for instance, on a day when he visited his sister, he also records seeing Mr. Taylor. Exactly which Mr. Taylor is being referred to isn't known, except that this one passed on some information about the Literary and Philosophical Society about them having been awarded their Diploma and Seal, and that they would send Moses his share. This was a relatively short-lived society that Moses contributed to in terms of papers given, but it had run out of steam before 1820, after a life of less than ten years.

Moses would read papers relating to astronomy to the Society, an example of which was during an extended Christmas break from his missionary. At a General Meeting of the Society on Monday 13th January 1812, he read a paper relating to a Comet. From September 21st 1811 Moses contributed articles to the Preston Journal relating to this Comet, each accompanied by an illustration of its progress and path. The series of twelve articles continued until December 14th 1811. The content of the paper can be read in Appendix Two.

For all the wrong reasons, probably the most eventful visit to Preston was in the October of 1811, when, as was often the case, he called to see Father Joseph Dunn, more often referred to as 'Daddy' Dunn, the Roman Catholic Priest at St. Wilfrid's. He would inform Father Dunn if he came across poor and sick people of the Catholic faith who were in need of relief or other assistance.

However, along with Isaac Wilcockson, the proprietor of the Preston Chronicle, Father Dunn was the co-founder of the Preston Gas Company, the company that were responsible for Preston becoming the first provincial town to have gas street-lighting. On one particular occasion, Moses was about to take his leave when 'Daddy' Dunn invited him to inspect a collection of bottles in which gas was being extracted from coal by a process of gradually increasing heat. Father Dunn described the resultant chemical as ammonia of coal. "Take it and smell it," Father Dunn said. Moses described it as strong and disagreeable. He offered him another one to smell, "This one is salts of coal," he said as he passed him the white coloured contents. About a minute after Moses had smelled it he was 'seized with a trembling and a palpitation of my heart. A cold sweat and a loss in a measure of my sight and sense. When I came to myself my pulse beat quick and then stopped. My throat felt very disagreeable and hot, and what I spit out of my throat into my handkerchief was effervescence'. Moses complained what hurt it had done his head, but the priest seemed surprised and said that it had previously cured several heads that had ached. Moses challenged him to smell it himself, and he appeared to do so. He decided that he had better get off, because it was now settling on his stomach and he thought he'd been poisoned. "As I went along the street it frothed up at my mouth like barm or suds. I called at a friend's to get some milk, and took some rhubarb. I threw up stuff that edged my teeth. I took a large quantity of salt and water. I threw up a large quantity, had a pain come in my left breast which descended till it got under my left rib. There it rested." The following day he commented that he could still feel the pain in his left side, and once again towards the end of the week

Although many of Moses' diary entries were just notes and, truncated ones at times, there were occasions where he went into great detail about his experiences. Possibly it was when his observations were at the extremes; of things that made him happy or completely the opposite. When he preached in Kirkham on the 17th February 1811, there were very few in the congregation in the afternoon, but he was impressed with a man called Kirby, a butcher. He commented on how he sang heartily, and how he seemed to have taken to their ways and manners. He made a point of speaking to him after the service and asked him to go again to the evening service. He went on to write, "He did, and brought more with him, and was hearty beyond description. He sang; showed people to their seats, snuffed candles, shut doors, and when they sang, he scolded James if he did not sing hard. He seemed all fire, and I was pleased to see it." Moses stayed in Kirkham that night, noting "how well he was entertained by that poor family." Poor or not, he noted his displeasure while he and some of the family were praying the following morning. He wrote, "I was astonished at the conduct of one of the sons, storming up and down demanding that someone bring his stockings. Nor did he stop until a little girl took them to him." It was unusual for him to make an adverse comment about incidents that happened where he was staying, so it had clearly offended him.

On a future visit to Kirkham he again made reference to the butcher, Kirby. On this occasion he wrote that he thought that Kirby had had too much to drink before arriving, "and was very windy". He went on to described how he, Kirby, and a man called James Parkinson sang together. Kirby sang tenor, and Moses sang bass, and afterwards Kirby had said, "Well done, I could do with you forever," and Moses replied, "I could do with you if you would be a good lad."

Moses spoke often about the difficulties he encountered when travelling from place to place. Near the River Wyre the banks were often breached and the roads covered with water. Indeed, when he was re-directed across neighbouring fields he was often prevented from doing so by the extensive flooding. He made an amusingly descriptive comment when walking along the narrow roads that criss-crossed the moss-land north of the river when he wrote, "We went calf-leg deep in the dirty moss roads."

During the final month of his missionary in May 1812, an incident that served as the catalyst for his ending it occurred in Garstang. Moses was in the house of Mr. Wright, who would seem to have been the man charged with over-seeing Moses' work. He saw, among some tracts on a table there, a large envelope, on the outside of which was written in large letters, 'Character'. He opened the envelope and found it to be a description of his own character given under six headings, and the analysis of his various abilities was not too flattering. This seemed to hurt Moses a great deal, and he finished as a preacher at the end of the same month.

On the 16th May 1812, after picking up the envelope referred to above, he wrote in his journal, "What is this thought? It is me. I took it up to read, but my conscience pricked me. I laid it down. I had no sooner done it than a thought struck me it was providence had put it in my way. I took it up and read it. It was under six headings:

1. Piety. This was genuine tho' mixed with some weaknesses.
2. Health and Constitution. Not the best. This was inferred from a little appetite.
3. Judgement. Sound with regards to the Fundamental Doctrines of Methodism.
4. Certainly has an investigating mind, but it was on the curiosities of astronomy more than Divinity.
5. Tho' abilities not brilliant, his way and delivery disgusting. Only had heard him once.
6. His behaviour as far as I know is unblamable.

He then continued to record, "I had scarce read it and copied the heads, when he returned." He asked me to go and look at a cow, which I did. He asked me 'What do you think it is worth?' I said, 'It may be worth twice as much or, not half as much. I don't do cashes'. He could see I wasn't in the humour to talk about cashes. He talked about me going to the District Meeting in a friendly way, but I did not say anything that I had found that wretched letter. I went into my room, knelt down before the Lord. I cried, and told him my woe, I think about an hour. Then I wrote this:

Dear Sir
If I did not believe you to be in a very good humour I would not give you this, but I do, and you know me, so let us be good humoured still. You talk of me going to the District. I declare that I am afraid of going, and this is a weakness, but the courageous,

when they know that a well-formed judgement is prepared to play on them, will make them to tremble, but to trifle no longer."

He then reiterated what he had said about his conscience pricking him, before taking up the letter once more.

He continued his letter, using the same headings,

1. Piety. Allowed, tho' mixed with some weakness.
   Observations: Should like to know them. Perhaps I might mend.
2. Health and Constitution. Not the best.
   Observations: Is this true? I durst not say it is.
3. Judgement sound.
   Observations: Query whether my constitution be not sounder.
4. Certainly he has an investigating mind, but it was on the curious more than the Divinity.
   Observations: This I would not say true if I said this.
5. His abilities not brilliant. His way and delivery disgusting.
   Observations: Is one time sufficient for judgement.
6. His behaviour unblamable.
   Observations: Thank you Sir, you deserve it.

He continued his journal by writing, "I now was called to supper. Mr. Wright was more than usually merry while sat at the table. I said to him, 'Sir, did you ever see an astronomical observation, and how it worked?' 'No' said he. I said, 'I will let you look at an observation and how it is worked, if you will promise two things:

1. To give it me again, and
2. Not to be vexed for it may not suit you.
   He promised both, and laid hold of it with eager hands.'

I have noted that Moses had dropped the word astronomical when he asked if he would like to look at an observation and how it worked. In reality it was his reply to the character comments referred to above.

"While he read it he changed colours", Moses wrote, continuing "He got up, threw it at me in such a rage as if he had a gouty toe and I had trod on it". I said, "Come, come, let us shake hands." And I would and did before I finished my supper. He said I had no business with his papers. I said he had no business with such lying papers, or to have such papers loose. The time of prayer came. He gave me the Bible. I opened it to one of the Psalms without desire that it might have been penned on purpose. I read it and burst into tears while I was reading. The same when I was praying. When we got up he stormed again as if he had been a sea-captain in a storm. He went to bed in this way." I suspect that, despite the request, Mr. Wright *was* indeed vexed.

From this point, only thirteen more days remained before Moses brought his missionary tour to an end. Events moved extremely swiftly following the disagreements with Mr. Wright, for just two days later, on Monday 18th May, 1812, Moses went to Preston and boarded a coach to Manchester. Whether this was the outcome of that fallout isn't clear, but he recorded that in the same coach was "the young preacher from Lancaster, Mr. Newton, and Thomas and Sam Jackson. Got to Manchester. Mr. Taylor sent a man with me to Mr. Yeats, corner of Newton Street, near Oldham Street Chapel, heard Mr. West preach."

The following morning he wrote, "Their meeting began, but I felt like a lost sheep. Mr. Ault came and told me that I should not be called on yet, so I went and read Buchanan's researches in Asia. Led Crosley class."

There are a number of queries raised in the quotes in the last two paragraphs, such as was Thomas Jackson the Superintendent of the Preston Methodist Circuit, and who was

Sam? Mr. Ault was a Preston acquaintance or friend of Moses, but why was he there? Who were Mr. Taylor and Mr. Yeats?

It was only on the following day, the 20th, that he was called, and all he recorded in his journal was "I went with the other young men and was examined, and passed the District Meeting." He then noted that he went, in the evening, to hear Bridgenorth preach at Salford. If the meeting had been a direct result of the conflict in Garstang, one can only assume that he gave an adequate explanation. There was no mention of Mr. Wright being there, and I think it was all due to the fact that the Fylde area was being removed from the jurisdiction of the Preston Circuit to one centred on Garstang. There's no doubt that Mr. Wright had a desire to succeed Moses in the Fylde, but I suspect that there were many people who would be less than pleased to lose him as their preacher. Indeed, when the new Methodist 'barn-like' chapel was built in Thornton in 1812, the preacher was the Rev. John Wright, still based in Garstang, and said to be 'covering a huge area as a travelling preacher'. Moses never referred to him as a Reverend, so I think he must have been ordained shortly after May 1812.

The day after, Moses went to Swinton where he preached in the new chapel, and the day after that to his hometown of Bolton where he called on his brother, although he doesn't identify which one. He then visited a few of his friends before travelling on to Hoghton Tower, sleeping at J. Harrison's, who is believed to have been the owner and occupier of Bolton Hall, which stood in the shadow of the Tower.

His final four days were spent successively in Preston, Freckleton, Bryning and Lytham, before finally returning to Poulton to make the necessary arrangements to draw the curtain down on an eventful eighteen months as an itinerant preacher.

Chapter Four

1811 – 1820 **The Post-missionary Years**

Although some work may have been carried out whilst he was still involved with his missionary work, the period immediately following it must have been largely consumed with work to prepare him for an extended series of lectures, which I feel certain must have been fermenting in his mind for several years. The majority of that work will have been the production of various items of equipment for use in the proposed lectures. There are unconfirmed suggestions that Moses gave lectures in his native Bolton in around 1806 - 1807 whilst he was living temporarily back there, but I would suggest that he had little equipment at that time, and that his lectures were of a different nature to the ones after 1815.

At the beginning of the 1800s a number of men were travelling the country offering similar lectures to the ones Moses proposed, such as those given in Preston in September 1812 by Mr. Handsford, his 'Astronomical Lectures – On Newtonian Principles' given in the large room at Mr. Whitehead's, the Bull Inn. There were five lectures at three shillings (15p) each, or ten shillings (50p) for them all. A further example can be found in Mr. Lloyd, who gave a course of four lectures at the Theatre Royal, Preston, in November 1823.

Much later, in 1852, the year when Moses delivered his final Triennial series which he called the Farewell Lectures, the Rev.

Doctor Daniel William Cahill, a six feet five inch Irishman gave a series of nine lectures on the same subject of astronomy, although the equipment used would seem to have been less flamboyant than Moses'. It would suggest that the appetite for this kind of event was almost insatiable in Preston, for the two men gave their courses just a month apart.

Quite unlike Moses' lectures, the Rev. Cahill's were held purposely to raise funds for the completion of St. Walburge's Church. Cahill (1796 – 1864) was an Irish preacher, lecturer and writer who travelled to the U.S.A. and Canada for those purposes, with most of his lectures being given free of any fee when they were for religious or charitable causes.

I am inclined to believe that once his missionary finished at the end of May 1812, Moses will have lost little time before preparing for what was to be the defining part of his life. In this decade, for instance, he constructed an orrery for his use in his lectures, together with several magic lanterns illuminated by limelight. For use with the lanterns, he purchased a collection of glass slides from a leading manufacturer in London, but was so disappointed and disillusioned with them that he painted and produced his own. He needed pictures of the planetary bodies as well as numerous diagrams and emblematic figures to project.

Of course, Moses would have to have provided some sort of income for himself during this transitional period, but there is nothing known as to whether he had any paid employment. In February 1814 he placed an advertisement in the *Preston Chronicle* offering to teach a limited number of pupils in geography, astronomy and the use of globes, two evenings each week, at ten shillings and sixpence (52½p) per quarter and five shillings (25p) entrance. The astronomical instruction was to be accompanied by practical observations with the telescope and quadrant. He further offered private instructions at their own houses to young ladies and gentlemen, in the same subjects at two shillings and sixpence (12½p) per lesson. The overall success of this venture isn't known, but he did teach the daughter of George Horrocks, and possibly the children of other members of that family.

Over the years, Moses proved himself, time and time again to be an ingenious practical mechanic. We have already seen how he was already adept at producing telescopes for both his own use and that of others, and was also capable of grinding and polishing his own lenses for use in those instruments. He also constructed microscopes, and other equipment, including a solar microscope. As I have said previously, there is a strong thought in the minds of the authors of the book, 'Robert Pateson: A Scientific Philosopher,' John Garlington and Emma Heslewood, (2004), that Moses Holden, a fellow member of the Institute for the Diffusion of Knowledge, and close neighbour of Pateson who lived in Bow Lane, had passed on his skill of grinding and polishing lenses for Pateson's own use and benefit as one of the earliest photographers in Preston.

People of note for whom Moses constructed telescopes were Sir Hesketh Fleetwood, the Rev. R. Carus Wilson, who claimed that the telescope he made for him was a superior instrument than one he had purchased from Peter Dolland and Company in London, the premier manufacturer at the time. There exists a letter from George Horrocks extolling the quality of lenses that Moses had ground and polished for use in his telescope, making the same comparison with Dolland. Dolland and Aitchison are the result of a merger between companies bearing the two single names, with Peter Dolland's company being founded in 1750. Aitchisons was founded in 1889 with the merger occurring in 1927.

Whilst this book was never intended to be a technical work, I think for the enjoyment of the reader it may be useful to learn, in layman's terms, what some of the words mean that we may come across when considering the early days of astronomical lecturing. I'll mention the orrery here, and then to other terms as and when we meet them.

An orrery is an instrument that most people will have seen, even if they didn't know what it was called. It is a piece of machinery, usually clockwork that is intended to demonstrate the movements of one planet with another. They can be as simple as one that demonstrates the relative movements of the Earth, Moon and Sun, or as complicated as one that demonstrates the movement of all the major planets in relation to the Sun. Moses Holden constructed one that included Venus, Mars, Jupiter, Saturn, Uranus, plus the Earth, Moon and the Sun. Mercury wasn't discovered until 1846.

Imagine for a moment the mathematical calculations involved in producing the various cogs, that when assembled will give precise movement to the various parts. The gearing for one planet will be totally different to the next, and yet Moses produced an instrument that was described by those competent to do so, to be unequalled in terms of its correctness.

It is true that he received some assistance in the production of the larger cogs. They were designed by Moses and made in the foundry of one of the Horrocks' mills with help from William Taylor and William Elsworth who both gave him their technical knowledge, and Edward Simons, a foreman mechanic, who assisted in the cutting of one of the largest wheels from patterns which Moses himself had produced. The smaller ones were said to have been made by Robert Westmore, a watch and clockmaker of 183, Friargate, Preston, who cut some of the finer-toothed wheels, but there must have been a host of varying sized cogs of uncertain provenance that went to make up the final instrument, which was assembled by Moses alone.

The Horrockses growing business empire was instrumental in Moses' advancement, but individually, the Horrocks family contributed also to that end. Moses gave lessons on astronomy to the daughters of Samuel Horrocks, M.P. for Preston, and he enjoyed a long-term friendship with George Horrocks, for whom it is known he produced equipment like a telescope, and lenses for telescopes he already possessed. Others who helped him were Thomas Norris of Howick House, and formerly of 'Redvales' near Bury, Isaac Wilcockson, proprietor and editor of the *Preston Chronicle* newspaper, Major Cross of 'Red Scar' in Ribbleton, and Sir Hesketh Fleetwood. There were doubtless many more whose offspring were the beneficiaries of Moses lessons in astronomy, geography, and mathematics.

There were some works available concerning the construction of an orrery, including 'The Description and Use of a New Portable Orrery: On a Most Simple Construction', by William Jones (1784), but I suspect that Moses would have considered himself capable of constructing something far more elaborate and complicated, and is likely to have used such a work only for the formulation of ideas.

Nothing is recorded regarding the short-term or long-term storage of the orrery, but I think that it would be on Horrocks's premises, which was most likely to be in the Spittal Moss, Fylde Road area. The finished size of the equipment has never been disclosed, but advertisements for his lectures quote a 'transparent orrery, twenty one feet in diameter'. Indeed there are some that say twenty four feet diameter. The variation I think, lies in the word 'transparent', and that what was happening during his lectures was that the orrery was back-lit by the magic lanterns, with silhouetted images being projected on to a screen. Some screens may have been slightly larger than others, thus allowing for a slightly larger image to be projected. An advertisement for a series of lectures in Manchester claimed an image forty feet in diameter, but that was quite exceptional.

I have already mentioned that he constructed a magic lantern, and the thought has always been that this and the orrery were built more or less simultaneously for use together. The problem with that theory is that limelight wasn't invented until the 1820s, and prior to that there was only the far less bright lamp designed by Aimé Argand from around 1780 and not ideally suitable for Moses' purpose. However, without any evidence it has to be assumed that he managed with this less effective method.

The Magic Lantern, or Laterna Magica, was the direct ancestor of the motion picture projector, and could be used to project moving images, using various types of mechanical slides. Their original invention still remains a matter of debate, but they may have a history as far back as the 15th century in Vienna, and an engineer called Giovanni Fontana.

We will also come across unusual names such as 'Ouranologia', 'Geastrodiaphanic', 'Dioastrodoxon', and 'Phantasmagoria', and I think it will suffice to say for our purposes that they all are referring to the same type of entertainment, that supplied by the transparent orrery and magic lanterns. Perhaps in the later Georgian years when it was most used, the glamorised names helped to sell tickets for the event?

It may, at this point be of interest to learn that when Moses died in 1864, his family were offered, by whom we know not, a staggering £1,000 for the orrery and attendant apparatus, and the portable organ that he used in his performances. That sum was turned down, and a later offer of £140 was similarly refused. Following the death of daughter, Annie Leonora in 1881, along with other possessions it was auctioned off for forty shillings. The fashion, it would seem, was over. It was bought by Messrs. Arkwright and Latham, a pair of auctioneers who may have been acting on behalf of one of the Horrocks' family, for it has been suggested that it ended up in the possession of one of that family.

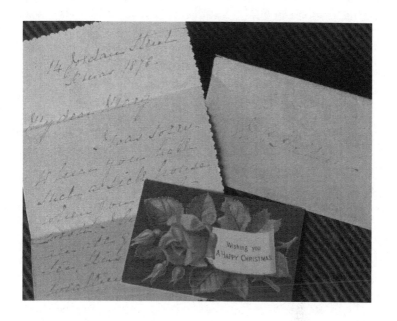

Letter and Christmas Card from Annie Leonora.

The Christmas card, which is the size of a standard business card, was enclosed with a letter written at Christmas 1878 by Annie Leonora Holden to her sister-in-law, Mary, née Blinkhorn. She was the widow of William Archimedes Holden, and the mother of his four children.

This may be the place to take a look at what a typical performance or lecture may have looked like. From a description at the time we read, 'the orrery that he constructed was a truly novel and ingenious instrument. Several magic lanterns containing painted slides and fitted up with revolving and other internal machinery were so arranged as to be completely under the control of his assistant. When Moses gave the signal, the lights in the theatre or lecture-room were covered, and the house placed in darkness. Shortly afterwards, to the solemn strains of a chamber organ, manipulated by Moses himself, the green curtain rose, and disclosed on a transparent screen a representation of the solar system, with the Sun rotating in space on its axis, and emitting powerful rays of light with the Earth revolving on its axis accompanied by its attendant Moon. The machinery was of such intricate character, that the earth and moon received their light apparently from the sun. The causes of day and night, the moon's changes, the action of the tides, and other celestial phenomena, were illustrated with wonderful clarity and realism'. What a marvellous picture that description portrays.

During his long tour after 1815, Moses gained the acquaintance of many scientific gentlemen, mostly better qualified despite possibly being less competent, whose friendship he retained throughout his life. His lectures were said to have been attended by all classes in society, and in almost every town shown the most courteous attention and hospitality from the gentry. Among those scientific men was William Rogerson of the Royal Observatory in Greenwich. Although there is no record of when they first became acquainted, it is said that he 'at a very early stage realized and appreciated Moses' talents, and the value of his orrery as an educational instrument'. He wrote the following letter to Moses on the 1st May 1833, and allowed him to publish it on flyers advertising future lectures.

When you exhibited your Transparent Orrery, I was indeed highly delighted. The appearance of the moon and planets were so natural, that I imagined myself looking through a powerful telescope: also the appearance of solvable and insolvable nebulæ through optic tubes. &c. In fact, the whole of your grand exhibitions, with the simple and clear instructions you conveyed by your lectures, left upon my mind sensations which I can never forget; and though I saw your scenery and heard your descriptions repeated at Hull, and in other parts of the kingdom where you kindly invited me to visit you, I was never weary of their repetition; and would deem it an indescribable pleasure, if I were so situated, to visit Preston at the present time, and witness the whole once more. Yes, my dear friend, although I am every day in the first observatory in the world, and using instruments of the very best kind, and viewing by them the belts of Jupiter, the ring of Saturn, the frozen poles of Mars, the horns of Venus, the spots of the Sun, and the Moon's varied disc, "rivers and mountains on her spotted globe," yet I should deem it a mental feast to see and hear your praiseworthy performances.

Signed: William Rogerson   May 1st 1833

Imagine also, the date of 1815 when this equipment was completed and ready for use after fifteen months endeavour. The railways were still a good twenty years distant, but somehow this equipment was moved from one lecture theatre to another. Just how big was it? How much other equipment was used in the lectures? Could it be dismantled in order to effect easier transportation? How long was he away on each visit, and how much extra luggage would that entail? Did he have any other option than horse and cart to move the equipment?

There is no documented evidence, so as an introduction into discovering how he managed to move his lecturing equipment as well as any other day to day or week to week luggage, let's take a look at some of the places where we know he lectured. Let us also make the assumption that Isabella, his wife, and slightly later William Archimedes, may have accompanied him, meaning an even greater quantity of luggage. He lectured in both Liverpool and Lancaster on a number of occasions, Leeds, Nottingham, Gainsborough, Hull, York, Nantwich, Chester, Manchester, Blackburn and Ormskirk. Add to that some other known facts such as the birth of William Archimedes in Pontefract in the same year, 1816, as he lectured in Leeds, and Banbury, Oxfordshire, where his second son, John Horatio was baptised in 1820. The common denominator that connects all the places mentioned is the canal system. That is not to say that he only ever used the canals, but it is a compelling thought, and a means of transport he used on his missionary tour when moving from Garstang or Bilsborrow to Salwick in 1811 and 1812. Perhaps those short journeys on what was a new form of transport sowed the seed for what, I am convinced, was to become an integral part of his plans for the future.

He also lectured in Whitehaven, the birth place of his wife, and there is no reason to believe that if water was his preferred means of travel, he didn't take a coastal packet boat from Preston to Whitehaven; and it may well have served as an alternative to the canal system for accessing Liverpool. Similarly, in 1825, he gave his series of lectures in the theatre in Carlisle, and it is interesting to note that Carlisle had a canal, constructed in 1823, to link Port Carlisle, formerly known as Fisher's Cross on the Solway, to the basin in Carlisle. After 1828 he continued to give his usual three night series of lectures, but it would appear, in the main, that he was a little less ambitious in terms of the distances travelled.

Theatre Royal, Preston

In 1815, when Moses Holden gave his first lecture here, the building was only thirteen years old, having been built especially for the Preston Guild of 1802. It was later transformed into a far more elaborate home of many forms of entertainment. Signor Paganini, the Italian violinist and composer appeared here in 1833 ... and promptly eloped with the pianist's daughter!

When he later referred to his tour of Northern England, he referred to it as continuing from 1815 until 1828. It may have been a date that he'd always planned to be the extent of the tour, but in that year John Horatio died aged seven or eight years of age. If they had been living and travelling on the canals, perhaps the death of the child precipitated the termination of it. There is no evidence to indicate the cause of death of John Horatio at the end of April, or whether it had any relevance to him ending the tour in that year. It had only been on the 3rd March, daughter, Annie Leonora made her appearance into the world, making the Holden's a five-member family for just a short fifty-eight days.

Coincidentally, Jordan Street, Preston, was built in 1827 and 1828, and I believe that Moses' efforts provided the means by which he could purchase a property there. I have seen four different numbers given as his address in that street, but they were numerical reorganisations by the Corporation rather than the family moving to another address. Jordan Street ran off Fishergate Hill by the side of the County Offices, and between Pitt Street and Bow Lane.

Prior to the construction of the County Offices, Jordan Street was the first fashionable street off Fishergate Hill to be developed as demand was stimulated for good quality housing as the cotton trade developed, and as well as being downwind and a suitable distance from the mills, it was equally a respectable distance from the unhealthy courts and alleyways that typified the town centre. Both the Fishergate Hill area and the Avenham area that was being developed at the same time, benefitted from wonderful views of the Ribble valley and estuary. The houses in Jordan Street were described as substantially constructed, with large gardens and for well-to-do people. It was around this time that Moses Holden began describing himself as a 'gentleman'.

In about 1835, Christ Church, a building with a strikingly Norman appearance, was erected at the end of Jordan Street and at the side of Bow Lane. In his privately published book about Preston's street names and their derivation, John Bannister singled out Tamar Street and Jordan Street as the only streets in Preston, with the exception of the Ribble, given the names of rivers. Despite efforts to discover the reason behind the name, I am left struggling to escape from the thought that Moses Holden had something to do with the naming of it. He had friends in high places, and was only six or seven years away from having the Freedom of Preston bestowed on him, and for whom else would the River Jordan have so much significance other than a devout Christian and Methodist preacher? Indeed, I noted with some interest that in July 2015, our new arrival, Princess Charlotte was baptised with water procured from the River Jordan. I'm sure Moses would have approved!

The orrery acquired its name when a complicated instrument called a Copernican planetarium was constructed for Charles Boyle (1676 – 1731), the fourth Earl of Orrery in Ireland, and the earl's title has become synonymous with all instruments of this kind. Copernican, meaning 'sun-centred', related to Nicolaus Copernicus (1473 – 1543) who was the mathematician and astronomer who formulated a model of our universe which placed the Sun at the centre of it, rather than the previous notion that the Earth was at its centre.

His first son's middle name was perhaps, an unsurprising choice for a middle name, for Archimedes, who died in the year 212 BC., was the man who was said to have possessed a primitive planetarium that was capable of predicting the movements of the planets one with another.

Before moving any further, let's take a look at what else he made. He constructed a solar microscope, which in reality is similar to a magic lantern, except that it is illuminated by the Sun rather than lime-light. Moses had one at his home, 'which he displays every day that the Sun shines except Sunday'. It was said to magnify 1,025 times. In his own words, he said of his solar microscope, 'that this instrument exhibits the wonderful works of God, in the minutiae of creation, displaying the organisation of vegetation, crystallisation, &c., and also a variety of living subjects such as eels, in paste or vinegar, cheese, mites, animalculae, &c.' Long after his death, the same or a similar instrument was in use at the Institute building opposite the Avenham Colonnade, and prior to that it had been in use in the Institute's former home in Cannon Street. Whether it was the same one as the one at his home isn't known, but doubtless he made more than one.

The description of Moses as a lecturer on astronomy would appear to be the diametric opposite of his performances as a Methodist preacher. From the strong, deep voiced, arm waving, Bible thumping messenger of the Gospel, his theatre performances were described as clear and relaxed. The former he saw as his duty, the latter was the labour of the love of his life.

A contemporary description of his lectures read, 'The settings were both novel, and for the time, spectacular; Moses played solemn music on a chamber organ prior to the commencement, and at a given signal all the lights were covered, with the exception of the ones attached to the orrery and lantern. The green curtain of the theatre rose, and disclosed on a transparent screen a representation of the solar system, with the sun rotating in space on its axis, and emitting powerful rays of light, with the earth shown revolving on its axis accompanied by its attendant moon.'

The first series of lectures using his new orrery were at the Theatre Royal in Preston. His astronomical lectures were always in three parts, delivered mainly on consecutive evenings, with a fourth night offered as a condensed résumé, for the benefit of the working classes, with a charge of sixpence (2½p) as opposed to the seven shillings and sixpence (37½p) and five shillings (25p) demanded from the upper and middle classes, a charge that permitted entry for all three lectures. It was more usual than not for his performances to be sold out, a fact which meant that by 1828 he was able to return to Lancashire and clear all the debts and other liabilities that he had incurred during his preliminary studies and the construction of his orrery and other lecturing equipment.

Like most new ventures, matters are often overtaken by events and unforeseen circumstances. Such was the case with the 1815 Preston lectures. The original plan was to hold them in late March, but Moses was obliged to rearrange them. In an advertisement in the *Preston Chronicle* on the 25th of that month, he announced that 'owing to the absence, during the Assizes, of many professional Gentlemen who have obligingly patronized his undertaking, he has been induced to postpone his lectures to Friday 7th, Monday 10th, and Tuesday 11th April'. This clearly indicates the type and class of audience that Moses was attempting to attract.

Just fifteen days later, Moses commenced the same series of lectures in the theatre in Lancaster, appealing to the same type of audience for patronage. This was probably the Grand Theatre which has a history in the social and cultural life of the city back to 1782, and is situated on St. Leonardsgate, just a stone's throw from the Lancaster to Preston canal. On this occasion he gave the lectures on the Wednesday and Friday of one week, followed by the third one on the following Monday. It was unusual for him to give his lectures with a pause between all the days, presumably because the equipment and scenery would hinder any other use; I'm sure there was a good reason.

It is most unlikely that I have identified all the places where Moses lectured, in fact, I'm sure I haven't, but in mid-July of 1815, he gave his three lectures at the Theatre Royal, Chester. Again, he spaced the talks in the same way as he had done in

Lancaster, so perhaps it was deliberate. Later in his career he more often than not presented them on successive nights, and if he was giving a recapitulation lecture on a fourth night, he would often leave a day free before presenting that one. The cost of a box seat here was nine shillings (45p) for the course, which was somewhat more expensive than the majority of theatres. Whether there was less availability or it attracted a wealthier audience isn't known.

On the 24th August 1815, Moses, now aged 38 years, married Isabella Spedding, a native of Whitehaven, and aged about 23 years. It is interesting to note that Moses had not yet lectured in Whitehaven, so is unlikely to have met her in that town. He married her in Liverpool, and I am persuaded to think that like a lot of Cumbrian girls in the early 1800s, she took a coastal steamer in search of domestic work in the growing port of Liverpool. I have never discovered why Moses had any particular link to Liverpool, but it was the birthplace of his hero, Jeremiah Horrox, of whom we will hear more in the next decade, but that would appear to be an insufficient reason. Perhaps he visited the city to attend sermons that would never reach the provinces; maybe he was seeking advice or materials in connection with his instrument construction, or even paving the way and making the necessary contacts with those who could make his lectures become a reality. Incidentally, 'Horrox' is said to be a Latinised form of 'Horrocks' and a flexible approach to its spelling is quite in order. Indeed, there are some who consider that there is a distant family link between Jeremiah and the Cotton Kings. Jeremiah was dead by the age of twenty-two, from an unknown cause, with one school of thought being that he perished during the first seventeenth century Civil War. But January 3rd 1841, seems a little early for that possibility.

The first of many courses of lectures in Liverpool did became a reality on Friday 11th August, 1815, followed by the other two on the 14th and 15th, the final one just nine days before his wedding. The lectures were delivered in the Great Hall in Marble Street, Liverpool, once the home of the Liverpool Debating Society and much more, and his wedding took place at St. Anne's Church, Richmond Street. Moses often related, with considerable humour, an anecdote relating to a 'professional seer' who cast his horoscope, 'that the stars declared that he would never take a wife,' to which Moses had responded by saying that he thought his own inclination and that of some young woman, who might happen to be of his mind, would have more to do with his matrimonial destiny than all the stars in the heavens!

There is an interesting and puzzling aside to the detail of where Moses married Isabella. Staff at the Liverpool Record Office, responding to my enquiry, informed me that there were two marriages recorded for Moses Holden, Astronomer. The second one was said to have taken place in the same church, but three years later on Christmas Day 1818, when it was alleged that he married a Mary Anne Coventry. It crossed my mind that he must have had an extremely plausible excuse for absenting himself on Christmas Day, but I felt obliged to accept it at face value.

Some years later, and somewhat by chance, I discovered a marriage certificate for Ms. Coventry, on the Christmas Day in question. She had married a man by the name of Owen Edwards, who was described on the certificate as a 'mathematical instrument maker', and in view of that I would be astounded to learn that Moses didn't know him. How the confusion arose isn't known, but I feel certain there's a link to Moses somewhere. Perhaps Owen Edwards was one of the contacts that I referred to in a previous paragraph.

On the 5th February 1816, a notice in the *Leeds Intelligencer* announced that it was Moses' intention to deliver a course of three lectures at the Theatre, Leeds, "as soon as fifty subscribers are obtained". I can find no evidence that the lectures materialised, but there is evidence that the family were still in the area in June of that year. First son, William Archimedes, was born in Pontefract on the 11th of that month, but not baptized until April 1818, by which time they were back in Preston.

I have no reason to believe that sufficient interest couldn't be raised in Leeds, for in 1817 he gave a series of lectures on the 1st, 2nd and 3rd of September, in the Theatre Royal in Nottingham. Advertisements for that event described it as "Illustrated with the Most Beautiful Eidouranion" or Grand Transparent Orrery. Although it is a somewhat circuitous route from Pontefract to Nottingham, including a stretch on the River Trent, I am somewhat encouraged by the fact that also along that same length of canal, Moses gave his lectures in Gainsborough. Thomas Cooper, in his book 'The Leicester Chartist', wrote, 'The visit to Gainsborough of Moses Holden, of Preston, to deliver lectures on astronomy, was a memorable event to me. I cannot remember who gave me the sixpence which enabled me to hear the first lecture; but I recollect that I drew out the figures of the zodiacal constellations, and of the solar system, and coloured them from memory, from the exhibition of Mr. Holden's orrery. I went round the neighbourhood, and showed them, and obtained pennies aplenty to enable me to hear the remaining lectures. This was in my twelfth year'. That detail would fit perfectly with Cooper's birth year of 1805.

I have since discovered that Thomas Cooper was a surprisingly similar character to Moses, although I am unable to say whether they ever met. It's unlikely, but I hope they did. He was a self-driven personality who had a passion for poetry, and who opened, at the age of twenty-three, a school. When not in the school, or following his trade as a shoemaker, he could be found conducting a role as a non-ordained Wesleyan preacher in the villages around Gainsborough. He soon fell out with his superiors in the church, when he accused them of not working as hard as he did. Later he became involved in the Mechanic's Institute in Leicester, the same organisation that in Preston became the Institute for the Diffusion of Knowledge, you may recall. It all sounds so familiar to me, and perhaps, in another life I'll return to take a closer look at Mr. Cooper of Leicester.

In mid-November 1817, Moses put an advertisement in the *Preston Chronicle* that read in a similar way to the one in Leeds, except that this one said he would deliver his lectures in Preston, 'as soon as sixty subscribers are obtained'. The likelihood is that those lectures were the ones delivered in 1818, as they were intended to be every third year until 1852, becoming known as the Triennial Lectures.

How the 1815 Preston lectures fared isn't known, so it is particularly interesting to note that on Saturday 7th March 1818, in the correspondence section of the *Preston Chronicle,* the following appeared, signed by 'Telescope'. This has to come from the pen of Moses, or from that of an associate at the behest of Moses, in order to grow his audience for his lectures just three days hence.

To the Editor of the *Preston Chronicle*. "Astronomy is probably the eldest, and unquestionably the grandest of all the sciences. It tends to enlarge and ennoble the mind, and to produce the most solemn and reverent ideas of the Maker of the universe.

O! How magnificent is the vaulted arch of heaven! Bespangled with innumerable orient stars! Thousands and tens of thousands of suns ranged around us, surrounded by ten thousand time ten thousand worlds, all in rapid motion; yet calm, regular, harmonious, invariably keeping the paths prescribed them, and doubtless peopled with myriads of intelligent beings formed for endless progression in perfection and felicity! This being the case, the greatest encouragement ought to be given to astronomers; they widen the empire of creation far beyond the limits, which were formerly assigned it. They give us to see that the Sun, enthroned in the centre of its planetary system, gives light, and warmth, and the vicissitudes of seasons, to an extent of surface several hundred times greater than that of the earth which we inhabit. They lay open to us a number of worlds, rolling in their respective circles around their vast luminaries. They go further, they let us know, that though this mighty earth and all our solar system, with all their myriads of people, were to sink into annihilation, it would be as much unknown and unnoticed by some other worlds, as the disappearance of a little star would be to us.

But I forbear saying more, because one of the best practical astronomers in this Kingdom will give a course of lectures upon the sublime science, in the Theatre, on Monday, Tuesday, and Thursday next, and as his apparatus is the largest, grandest and most accurate of any in the Kingdom. I trust he will meet with the greatest encouragement, especially from the respectable young people of this large and respectable town."

In the advertisement for the lectures he mentions for the first time that his new publication, the 'Atlas of the Heavens' is being offered for sale. A fortnight later there was a further notice in the *Chronicle*, announcing that on Easter Monday evening, the 23rd March, his 'Grand Transparent Orrery and all the splendid scenery' that had been exhibited during the course of lectures would be displayed for the inspection of the public, and that he would offer a short recapitulation of his lectures during the evening. Box seats were three shillings (15p), whilst the Pit and Gallery were two shillings (10p) and one shilling. (5p) respectively, and the 'Small Celestial Atlas', or 'Map of the Visible Heavens', were available at three shillings and sixpence. (17½p).

At the end of June 1818, lectures were delivered in the Minor or Royal Minor Theatre in Spring Gardens, Manchester. The advertisement in the *Manchester Mercury* on the 2nd June explained that the lectures would be accompanied by his beautiful transparent orrery, and adding 'without doubt the largest in the world', and also 'an immense quantity of splendid scenery'.

In November of the same year, 1818, Moses gave his series of lectures at the theatre in the beautiful ancient market town of Nantwich, through which runs the Shropshire Union Canal. A few days later the following appeared in the *Chester Courant,* "We hear that the inhabitants of Nantwich were highly gratified by a Course of Lectures, on astronomy, delivered last week by a Mr. M. Holden. Mr. Holden's abilities as an astronomer cannot but rank superior to Dr. Herschell's, having, as he informed his audience, telescopes of greater power than the Doctor's, being made by himself. Mr. H. also asserted that the Doctor's imagination of having seen a burning mountain in the Moon might be nothing more than a carpenter having set fire to a bundle of shavings! Mr. Holden has, no doubt made this wonderful discovery, through the superior power of his own telescopes!"

A SMALL

Celestial Atlas,

OR

MAPS OF THE VISIBLE HEAVENS;

Designed as a useful Companion for the

YOUNG STUDENT,

AS WELL AS FOR

THE PRACTICAL ASTRONOMER, &c.

By M. HOLDEN,

LECTURER ON ASTRONOMY.

'Οι ουρανοι διηγουνται δοξαν Θευ. Psalm, xix. 1.

Preston:

PRINTED FOR THE AUTHOR, BY I. WILCOCKSON.

1818.

The Small Celestial Atlas

No one was more acutely aware of the price of text books on any scientific subject than Moses, and that it created a barrier to the working man and the less well off. Not everyone was as assiduous as Moses, and, as he demonstrated later in his life, he was always keen to promote the education of that particular strand of society. It is true that people had to provide a living for themselves and their families, but in the early days Moses had been no different. As we have already seen Moses seemed to have the drive and ability to create the time, 'to persevere whilst other slept'.

How long the idea of the Celestial Atlas had been forming in his mind isn't known, but in the introduction to the first issue in 1818 he wrote, 'In travelling through a great part of the kingdom, I have been applied to by the conductors of many respectable schools, and likewise visited by private individuals, after I had exhibited the constellations belonging to my last lecture, who wished to know whether there were maps of the Heavens to be had. I informed them of Flamsteed's, a work at three pounds five shillings (£3.25p), and some others that were very much inferior, yet high in price. He also mentioned 'that excellent work', Francis Wollaston's Catalogue of the Stars and Nebulae, 'from which persons of ordinary ability might make their own (maps)'.

He goes on to tell the story of when he was a child, but adds the following detail which, I think for the first time, attributes an age to him. He talked about when he was younger being transfixed by the general gaze of the stars, their difference in apparent size and brightness, and their apparent difference in colour. He concentrated on no particular star, but 'the vast whole began to swell my breast, and burst upon my sight, in turns, in majestic grandeur'. He then went on to relate that, 'late one evening my attention was fixed by a very large, bright, and twinkling star, in the south east, which seemed to emit a variety of colours; I sought it another night, and found it south east of three stars, in an oblique line, and so on for months'. He then lost sight of it, but he found it again the following year, and in the place he had seen it the previous year. 'I began to examine the few works I then possessed, and by much labour, and with the help of a few tables, and a small catalogue of stars, I found this to be Sirius, the Great Dog Star'.

He then, making reference to his Atlas wrote, 'had such a work as this been then in my possession, I should have known, by bare inspection, that this was Sirius, and also that the three stars in an oblique line were part of Orion. Nay, further, I should very soon, though hardly ten years of age, have known, with certainty, that these stars had names given to them by scientific men'. So he'd only just reached double figures, and in years, still a boy.

And so, using Wollaston's catalogue, referred to earlier, he was able to draw the twenty maps included in the Atlas, condensing a lot of information in a small space, including the magnitude, or relative brightness, of the principal stars, and their names using the Greek letters familiar to astronomy students. Each map has copious notes to accompany it, enabling the student to make full use of them, and each map includes a sign which indicates the pole star, informing the student which way the map should be held; rather like the arrow on a terrestrial map which indicates the direction of the North Pole.

The Atlas was reprinted on at least four occasions, the final one being in 1836.

Correspondence to the Newspapers

There are only about four pieces of correspondence available for this decade. There are probably others that remain undiscovered, but other things, as we have seen, were happening at the time. One of these letters, dated February 1818, gave his address as Friargate, but the actual site and details of his length of stay there, are yet un-established. You may recall from chapter one that he had lived 'on the edge of the town'.

That letter spoke of the spots that could currently be observed crossing the face of the Sun from east to west. That is from left to right, for everything in space is a reversal of what we are accustomed to on Earth. He described the position and the colour of the spots, and the shades of the margins that surrounded them, adding that it was as fine a range of spots on the Sun that he had seen for two years. Whether he wrote about that event is not known, but he did write about another notable sighting of spots two years before that in 1814.

An anonymous editorial piece appeared in the *Preston Chronicle* of 16th November 1816, relating to an eclipse of the Sun, but I retained it despite its many references to places in Yorkshire. The general phraseology sounded 'very Moses'. Later, I discovered that in that year he certainly spent time in Yorkshire, lecturing in Leeds and his son being born in Pontefract, so that I would now be astonished if he wasn't the author of it.

The Approaching Solar Eclipse

'The Eclipse of the Sun on Tuesday morning, the 19th inst., will be the greatest which has happened in this country for thirty-two years. At York, 10 digits, or ten parts out of twelve of the Sun will be eclipsed; at Scarborough and Whitby the Eclipse will be somewhat greater; and in a part of the Baltic, and countries to the north-east of us, the Eclipse will be central and total. To the south and west, it will be less than with us. At London, nine digits and two-fifths will be eclipsed. The duration &c., of the Eclipse at York, according to true time (by which our clocks are regulated) will be as below:

Sunrise, 23 minutes before eight o'clock.

Beginning of the Eclipse, eight o'clock.

Greatest obscuration, 5 minutes past nine o'clock.

End, 15 minutes past ten o'clock.

The Moon makes the first impression on the right of the Sun's upper limb. Those who adjust their clocks to apparent time, must observe, that the Sun being fourteen minutes before true time, the different observations by their clocks will happen fourteen minutes later than above stated.'

Another letter which appeared in the *Preston Chronicle* on the 17th July 1819 was signed by Moses, but it remained a mystery because he gave his address as Broseley, Shropshire. I later discovered that he gave his series of lectures in Nantwich in November of the previous year, making a visit to Broseley a lesser surprise. Broseley doesn't appear to have a canal link, but of course, for his astronomy purposes, he would have less to convey. Perhaps one day we'll discover where and with whom he was staying.

The letter itself related to a Comet that had been visible for two weeks, and was predicted to be so for several months. He said that it would be visible using a perspective glass, an early form of telescope; or a day and night glass with a magnification of sixteen or twenty times. To put this into context, a modern field telescope, beloved by birdwatchers, will have a lower magnification of more than twenty times, up to sixty or eighty times, with those used by astronomers being several hundred times upwards.

Returning to his lecture tour, the final years of this decade are strangely under-recorded. There is an absence of information about any activity, whether that was his astronomy, lecturing, correspondence, or his delivering of sermons. The only clue that I have comes right at the end of the decade, on the 29th September 1820, and I only became aware of this when, after twelve years research I was contacted by a man who announced that he was Moses Holden's great, great, great-grandson. He was the first living descendant of my subject I had encountered, and he lived in Hertfordshire. He provided me with a family tree that, while still not entire, included details of another son, a brother to William Archimedes. Here now, was a boy with an equally impressive name, John Horatio, where the family tree I just referred to, informed me that on the date mentioned above, he was baptised in Banbury, Oxfordshire.

Banbury, Oxfordshire? Moses is only on a lecturing tour of the northern towns and cities, I was informed. So what's he doing in Banbury? Is he away for such a lengthy period that he can't wait until he returns to Preston to have the child baptised, as he had done with William Archimedes? John Horatio was dead by 1828, so perhaps he was a sickly child, and baptised elsewhere in the event of him not surviving. I must conclude, therefore, that he was away from Preston for an extended period; but doing what? Moses wasn't a man who ever did nothing.

Banbury, it will not now come as a surprise is on the canal network, on the Oxford Canal, not very far north of the University City. Did he ever visit that place? So many questions, but apart from the baptism, I have nothing to relate between his lectures in Nantwich in 1818, and his lectures in York in September 1824.

How many secrets remain to be discovered?

Chapter Five

1821 – 1830

Despite having indicated that Moses gave lectures every three years in Preston after 1815, there is no record of any in 1821 and 1824. Indeed, we will discover in 1827 it was said that he hadn't lectured in Preston for nine years. His whereabouts from 1818 until 1824 are only scantily recorded, but from the little I know, I suspect he was in Yorkshire, south certainly as far as Banbury in Oxfordshire, and possibly further.

The first knowledge I have of Moses after his Nantwich date, is on the opposite side of the country in Lincolnshire. On Thursday 18th September 1823, I found him attending the annual meeting of the Horncastle Circuit of the Wesleyan Methodist Missionary Branch Society at the Methodist Chapel in that place. In a lengthy editorial about the event it commented that despite being in the midst of the 'necessary bustle' of a late harvest, and a day when there 'were two or three markets and stock fairs in the area', the congregations for both sermons were large. Moses wasn't one of the preachers, but was well enough known to be singled out as an attendee, describing him as 'the astronomical lecturer'.

Horncastle has its own canal that runs due south to where it joins the River Witham in a region called Dogdyke, and from that point there is a plethora of navigable river, canals and navigable fenland drainage ditches or dykes, the first of which takes you right into the centre of Lincoln. The town of Horncastle lies about twenty miles to the east of Lincoln, and at the time a great deal of lobbying was being conducted to provide a memorial statue, preferably in the cathedral in Lincoln, to Sir Isaac Newton, a native of that county. The idea had been turned down by the Corporation some four years previously on matters of expense, but the cause rumbled on. There is no doubt that Newton will have had the admiration of Moses, so maybe he was lending his considerable voice to that cause.

The Theatre Royal in Lincoln was built in 1806 on the footprint of a previous theatre that had been there, on Butchery Street, since 1764. Its layout was identical to several other theatres that went by that name, and which Moses seemed to favour, and I think there is little doubt that he will have attempted to present his lectures there, even if he was unsuccessful.

From one county town we move northwards to another, York. There are no details of his appearance there other than his lectures had begun on Monday 11th September 1824. York also, has a Theatre Royal, that one having made its first appearance twenty years before the Lincoln Theatre. The Yorkshire Gazette carried a four line notice two days before the opening night, commenting 'the science is so interesting in all its details that we cannot doubt but the attendance will be numerous'.

Later the same year, the *Durham County Advertiser* of 4th December 1824, reported that 'On Sunday last, two excellent sermons were preached in the Methodist Chapel, Northallerton, by Mr. M. Holden, lecturer in astronomy, after which collections were made for the benefit of the Sunday School, amounting to six pounds eight shillings and two pence halfpenny (£6.41p). Today, that sounds a trivial amount, but when a decent weekly wage for the common man was 50p to 75p, then eight or twelve weeks wages was a not an inconsiderable contribution to one of Moses' favourite causes.

The penultimate month of 1825 saw Moses deliver his programme in the most northerly place that I have discovered him. I have seen mention of him visiting Scotland, but found no hard evidence. However, Carlisle is as close as you're likely to get without actually being there and there is plenty of evidence to support the visit. The question that has plagued me is how was he able to get there with all his equipment and other luggage?

I think a tortuous journey over Shap Fell in 1825 can be discounted, and once again I think we can look towards the water. However, the Preston to Kendal Canal would have been of little help, getting him barely to base-camp on Shap. It is more feasible that he travelled from Preston on a coastal sailing vessel, in the manner that I described in the last chapter.

Despite what must have been an additional effort, his visit to Carlisle was one of his least successful. By this time he was sufficiently experienced not to have offered the talks without the minimum amount of sponsorship offers being received, but the *Carlisle Patriot* reported that, *'the audiences were small – not at all equal to the merits of the lecturer and the excellence of his apparatus.'* It went on to describe Moses as *'a clever and most indefatigable lecturer, with a voice equal in compass to the largest building, and relates an anecdote with a pleasing humour and much force of expression; and in this point he is materially aided by a north country accent which imparts to all he advances an air of blunt sincerity more powerful than balanced periods, and an affected delivery.'*

It is probable that this sort of educational entertainment was regularly poorly received in Carlisle, for Moses said afterwards that despite the audience being more scanty than he would have hoped, he was still grateful, and would not say, as some had done, that he should refrain from visiting Carlisle again at a future date. I have no evidence, however, to say that he did.

In April 1826, Moses gave his series of lectures in Whitehaven, the home town and birthplace of his wife, Isabella.

Unlike his visit to Carlisle, it was noted in the press that the audiences were of a very respectable size. This was the year that he provided the finances for a memorial to his early inspiration, Jeremiah Horrox, and although that didn't happen until November 1826 he was clearly thinking about it in Whitehaven.

During the course of his lectures there he made reference to *'some wonderful calculations made by the Rev. J. Horrox, a clergyman of the Church of England, who resided at Hoole, about eight miles from Preston on the Liverpool Road, calculations which even at the present time could not be excelled for accuracy, yet Horrox had died in 1640 at the age of twenty two or three! To the talents and research of this young man the whole of the professors of this sublime science who have flourished since his death, both in England and on the Continent, are in no small degree indebted. Even Sir Isaac Newton himself may be said to have built his superstructure upon the foundation laid by this great genius; and to the disgrace of the country,'* Moses remarked, *'there was not even so much as a stone to mark the place where his remains were laid!'*

He went on to explain that when he returned to that part of the country (Liverpool) he proposed giving the receipts of a course of lectures to erect a monument to perpetuate his memory.

It should be pointed out that it is certain that Horrox was not a Reverend. At the most he was the Curate of St. Michael's, Hoole, at the time for which he is remembered, and we'll take a closer look at him now, because in November 1826, Moses returned to Liverpool, where on the 20th, 21st and 22nd, he delivered his course at the Pantheon in Church Street. The Pantheon had only been constructed in 1820 as the 'Dominion of Fancy', and was then reconstructed in 1824 and reopened in 1825 under the new name. It was said to have a small stage, and seating for around one thousand. The Pantheon was only the second theatre to be built in Liverpool, the previous one being the Theatre Royal, where Moses held his 1815 course of lectures.

Two days after that series, on the 24th November, he delivered the lecture intended to raise the money for Horrox's memorial, and in an article in the *Liverpool Mercury* of that date there was a note to say that, *'we have very great pleasure in stating that Mr. Holden's lectures at the Pantheon, are nightly received with the most lively approbation, and that on this evening, in addition to the recapitulatory lecture, Mr. Holden will give a complete description of the imminent solar eclipse'.*

The article also carried the timing of the eclipse, in the latitude and meridian of Liverpool, provided by Moses, for the benefit of potential observers:

Eclipse begins:	9 hr. 35 mins 26.55 secs.
Middle:	10 hr 40 mins 43½ secs.
End:	11 hr. 47 mins. 58½ secs.
Digits eclipsed:	6° 57' 52.56 secs.

At around the same date, the *Preston Chronicle* announced the same event, with their 'Poulton Correspondent' probably Moses, giving the relevant timings for Poulton. There it would begin at 10.10am, middle out at 11.15a.m, and finish at 12.23p.m. At the peak of its obscurity, the eastern part of the Moon's disc will pass on the Sun's western limb, when its appearance will be similar to that of the Moon when near her first quarter.

The cost of the memorial is unknown, but it remains on view in the Church of St. Michael's, Toxteth Park, although there seems to be a preference to name Aigburth as its position, and also to refer to it as St. Michael-in-the-Hamlet Church. It is what was referred to as an 'iron church', constructed in brick, but with many cast-iron components, including parapets, battlements and pinnacles, and the roofs are of slate slab set in a cast-iron framework. It is protected by an English Heritage Grade 1 listing.

St. Michael-in-the-Hamlet Church, Aigburth, is one of a number of iron-churches, this one built by the Mersey Iron Foundry in 1813 – 1815, although it has been renovated and enlarged since that time. The east window in the church is a stunning work of stained glass.

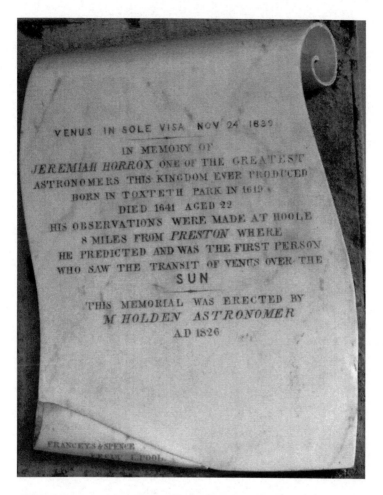

VENUS IN SOLE VISA NOV 24 1639

IN MEMORY OF
JEREMIAH HORROX ONE OF THE GREATEST
ASTRONOMERS THIS KINGDOM EVER PRODUCED
BORN IN TOXTETH PARK IN 1619
DIED 1641 AGED 22
HIS OBSERVATIONS WERE MADE AT HOOLE
8 MILES FROM *PRESTON* WHERE
HE PREDICTED AND WAS THE FIRST PERSON
WHO SAW THE TRANSIT OF VENUS OVER THE
SUN

THIS MEMORIAL WAS ERECTED BY
M HOLDEN ASTRONOMER
AD 1826

FRANCEYS & SPENCE
LIVERPOOL

The creators of the scroll, Franceys and Spence, were a Liverpool firm that operated until 1844, despite Samuel Franceys dying in 1829. It is described as a brown mural tablet, bearing a statuary scroll, with the representation of Venus passing the Sun's disc. The latter feature has now worn away with the passage of time.

I have recently discovered, in the Waller Manuscript Collection at the University of Uppsala, Sweden, *(Waller Ms alb-33:060a letter)*, a letter written by Moses Holden to Isaac Wilcockson, the proprietor of the *Preston Chronicle,* in which he indicates his intention that if permission were to be refused to erect his memorial in St. Michael's in Toxteth Park, then he would have it placed in the similarly named St. Michael's at Hoole, where the observation was made.

This discovery, together with a comparable one where local historian Heather Crook came across a book of poems, written by Walton-le-Dale poet, Henry Anderton (1808 – 1855). Anderton had been a follower of Moses as a local preacher, and would go to great lengths to listen to his sermons. Such was the impression that Moses had on him, he wrote several poems about him which were published in the book. Where was the book? In the library of the University of California in Los Angeles! See Appendix Six to read those poems.

I have for many years been driven by the thought that the only time that I stop finding things is when I stop looking, and with examples such as these, there is constant encouragement that there will be something else just around the next corner.

Towards the end of February 1827 Moses wrote to the *Preston Chronicle* with news of two events. Firstly the occurrence of spots on the face of the Sun, and instructions of how to observe them, but the more interesting part was relating to the planet Venus which he had observed close to the Moon at 11a.m on the 22nd February. He wrote, "I saw it with the naked eye very distinctly, but with the telescope, and a power of 130 times magnified, it looked larger than the Moon does to the naked eye. It was not half enlightened, but a large crescent. It is below the Morning Star, and might be found about 10a.m if the air is clear.

For those who may be interested, Venus is sometimes called Earth's 'Sister Planet' because of its similar size. Not only is it the closest planet in size to the Earth, but also the closest to it; but with an atmosphere consisting of ninety six per cent carbon dioxide there are a lot of differences as well.

***

In March 1827, a notice appeared in the *Preston Chronicle* indicating that Moses Holden had been *'spending some weeks of leisure during the winter in his native town'*. It is true that I can throw no light on how his winter was spent, but Moses never, as I observed earlier did anything that could be construed as wasting his time. We will discover his involvement with the creation of a Mechanics' Institute in 1828, and I see no reason why Moses won't have been looking at similar institutions in other towns and cities, in order to draw inspiration during the preceding months.

The first Mechanics' Institute to be opened in the country was in 1823 in Liverpool, a place that was familiar to, and often visited by, Moses Holden; the following year one was opened in Leeds, Manchester and Bolton; but more of that in due course.

The other part of the notice was devoted to the pending series of talks to be given in Preston, and is couched in language that suggests that all its readers will be familiar with Moses. *'Those who have a personal knowledge of the great merits which belong to this gentleman, and who are acquainted with the difficulties through which he has attained the high station he now holds in the line of his profession, will require nothing from us to urge them to give their active patronage on this occasion'*. It went on to comment that *'It is now twelve years since Mr. Holden commenced his career as a public lecturer in this town. Since that period he has visited nearly all the great towns in the kingdom; and by constant attention to his profession, has brought his orrery and illustrative scenery to be the most complete set of astronomical apparatus that was ever in the possession of any former lecturer.'* This is the only time I can recall a reference to 'all the great towns in the kingdom,' rather than it being restricted to northern towns and cities, but the evidence is proving difficult to locate.

The lectures took place on the 26th, 27th and 28th March, and unlike previous, and some future events the admission charges were for each evening rather than transferrable tickets for the course of lectures. That gave the potential for a significant increase in revenue, and the start time of 7.00p.m rather than the customary 7.30p.m, suggests that he was experimenting with a few details.

An article in the *Preston Chronicle* after the events reported that this was the third time that he had lectured on astronomy in Preston, and so we can be sure that the 1821 and 1824 dates were missed. It also commented that on the first night the Theatre was 'very fully attended', but because of the inclement weather on the other two nights, attendance then was not quite so large. It was followed up, on Monday 2nd April with a recapitulatory lecture, with the *Chronicle* adding that 'after this night no part of the astronomical lectures can be repeated, as the theatre is engaged for another purpose'.

The article also made reference to 'the occasional' lectures about optics that Moses had given. He went on to give two within about twelve months in Preston, the first at the Theatre Royal, and the second at the Cockpit at the rear of the Minster, or St. John's Parish Church as it was at the time. Interest in them was further fuelled after he gave a demonstration using his solar microscope. This was only a couple of weeks after his 1827 lectures, so where they were held I don't know. He did have such an instrument at his house in Jordan Street, but this date is precariously close to that street not yet having been constructed, and so I think that can be discounted.

A long article relating to the event in the *Preston Chronicle* of 21st April 1827, is extremely interesting, but gives not a clue as to the venue, other than to say that *'Mr. Holden gratified a numerous party of private friends, on Wednesday last, by exhibiting some of the wonders brought to light in the minute part of the animal, vegetable, and mineral kingdom, through the agency of a solar microscope, which magnified the objects placed before the lens, to the extent of more than two thousand million times their real bulk. Mites produced in decayed cheese, which were only just visible to the naked eye, were presented to the spectators as lively, active animals, of three feet in length'.* The article continued, *'. ... perhaps the most striking part of the exhibition was to witness the process of crystallisation on placing drops of water saturated with different salts within the focal point'.*

Again there were calls for him to deliver his optics lectures, but he explained that unless he could give them in a suitable place, with the facilities he would require to display them to their best advantage, he would not do so, and it was the first week of April the following year when he presented them at the Theatre Royal. It is difficult to determine what form these lectures took; the advertisement spoke of a 'great variety of transparent scenery', and the use of optical instruments, but it doesn't go into detail. One would have thought, for instance, that for a solar microscope to be of any value, the Sun would be required to provide the light. The demonstration of this instrument for the benefit of his friends had, of course, been one of the catalysts for the talks.

Lectures on optics had been offered for well over a hundred and fifty years, by Isaac Barrow and the renowned Sir Isaac Newton. The first named had followed on the work of the astronomer, Johannes Kepler, and his revolution in geometrical optics by creating a mathematical theory of optical imagery, and image formation in lenses and optical equipment. In turn, this led to an explanation of the optical system of the eye. They were, in fact, lectures about the theory of optics, of telescope design and construction, and the laws of refraction.

However, having given this course of lectures, Moses gave further ones the following year, the final one, again at the Theatre Royal. The advertisement described that 'on that occasion he will illustrate the science by an exhibition of the Optical Illusions of the Magic Lantern, Grand Phantasmagoria, &c.' It was a single evening performance with tickets ranging from Boxes three shillings (15p), Pit two shillings (10p), and Gallery one shilling (5p), but added that free tickets of admission to the gallery will be delivered to the Members of the Preston Institution.

In a much later article, the magic lanterns were described as 'several magic lanterns, containing painted slides, and fitted up with revolving and other internal machinery, and were so arranged so as to be completely under the control of his assistant'.

We haven't yet encountered the Preston Institution, because it was only formed in 1828, with Moses playing an integral part in it. In the meantime, perhaps readers would like to read a report of the lecture on optics that Moses delivered in the Cockpit on the 16th April 1829, and according to a newspaper report was specifically for the benefit of the Institution I've just mentioned. It was said in the report that *'there were very few members absent, and the whole seemed highly gratified'.* The astonishing powers of the microscope and telescope were given in glowing terms. The art of grinding and polishing glasses of different descriptions *'was explained in so simple a manner, that those whose genius might lead them to the attempt, could not fail of being successful.'* The full report can be seen in Appendix Four.

In the report it becomes clear that whilst Moses is on his travels, on, we are almost certain, the canal system, he makes reference to experiments he carried out on wildlife, or as he called it, animalcula. He speaks of visiting the Fens of Lincolnshire and Cambridgeshire, the latter place extending our knowledge of his travels even further. It also enhances our understanding of how he was occupying himself. In addition to these experiments he would, given the opportunity, preach wherever he was able. We'll see further examples of this as the year's progress. He was also researching the opportunities to present his lectures, promoting them, conducting them, to provide a living for himself and his family.

Following the Cockpit lecture an article appeared in the *Preston Chronicle* reporting that it had lasted about two hours, *'which being interspersed with lively anecdotes, the company seemed ready to chide hasty-footed time for parting them'*. What a lovely turn of phrase! It further reported that Moses had announced that it was his intention to give an astronomical lecture in the Theatre prior to leaving his native town on a tour of lectures on astronomy or optics.

This is an interesting comment, for as I indicated earlier, his tour of lectures were said to have lasted from 1815 until 1828, but now, having helped, as we will discover in the next chapter, to set up the Institute for the Diffusion of (Useful) Knowledge, he is now announcing that he is off on his travels once more. It was said that by 1828 he had earned sufficient money to pay off all the debts that he had incurred during his preliminary studies and the building of his orrery, and I suspect that it was at this point he acquired the property in Jordan Street, and the point, also, when Isabella stayed at home in preference to a life on the peaceful waterways of England.

Chapter Six

The Mechanic's Institute or Preston Institute for the Diffusion of Knowledge

Probably the most enduring thing with which Moses would ever involve himself was the creation of what in most towns and cities would have acquired the name Mechanic's Institute. I indicated earlier that I believe that Moses will have witnessed their introduction in Liverpool, Bolton and Manchester up to four years earlier, and been keen to draw up a blue-print based on the experiences of those three, and maybe more places, for a similar institution in Preston; education of the working man had always been high on Moses' list of priorities.

There was a close link at the time between the inspiration drawn from the anti-spirits associations in America in the 1820s, and the movements being set up for the education of the working man. There were several men who were involved with this institute and later the Temperance Movement and three of those were Moses Holden and the far better known Joseph Livesey, cheese-monger and newspaper proprietor, and Thomas Batty Addison, Preston Barrister and Recorder.

Letters to the Editor of the *Preston Chronicle* had started to appear as early as 1825 regarding the desirability of a Mechanic's Institute in Preston, almost all of which were written anonymously, and articles outlining the many benefits that would accrue from the formation of such an institute. Studies of the Arts and Sciences, History and Biography and so on were envisaged as possible study areas, together with a library for the people, which, with the combined efforts and financial input of a large number of people would render the cost for each member a relatively small one.

As late as the 23rd August 1828, a letter appeared in the *Preston Chronicle* querying why after reading several pieces of correspondence in that paper, each of which recommended the establishment of a Mechanic's Institute, why there was still no sign of one. There was even a suggestion that as usual, Preston was late in adopting the idea. At this point I feel I should offer the suggestion that, although the letter was published anonymously, the general tone, structure and phraseology in the letter, would suggest that it was penned by Moses Holden or the man with whom he had a close working relationship with, Joseph Livesey. The letter was signed with a letter 'L', and has been accepted as most likely submitted by Livesey.

The writer states that from observations that he had made in places where they had an institution, there were those that were doing well and those that were failing, and that it was his opinion that those that were the most successful were the ones that had departed from the original idea of being composed solely of mechanics, and were now constituted of enlightened persons of every grade and every profession in society.

He went on to say that under those sort of circumstances he thought such an institution could succeed, adding that he would call it *'The Society for Promoting Useful Knowledge'*, which was similar to the name given to Lord Broughams creation, the *'Society for the Diffusion of Useful Knowledge'*, the *'S.D.U.K.'*

He made the observation that although the interest and backing of the wealthier members of local society was vital, he commented that every effort should be made to avoid the society or institution being dominated by them. He suggested that subscriptions for membership should be kept as low as possible, suggesting *not less than* one shilling and a penny (5½p) per quarter, leaving it sufficiently open so that the wealthier could pay more than that amount. He added that other societies had not succeeded because of the failure to keep subscriptions affordable to the 'operative' working man.

The writer continued by saying that the wealthier members of society should be willing to contribute to such an institution, not only by subscribing, but by further gifts of either money, books or apparatus. Any capital accrued could then be used to rent rooms and create a library.

He concluded his letter by appealing for those gentlemen who had, from time to time suggested such an institution, to get in touch with him through the newspaper office, indicating their *'willingness to assist heartily'*, adding, *'attached as I am to Preston, for the people's sake, I am anxious that it should be second to none in really useful institutions'*.

The following Saturday, the 6th September, there was an equally enthusiastic and encouraging letter, reiterating much that was said in the previous one, and the week after that, a letter from a reader describing himself as *'a plain common man'* did much the same as the previous correspondent.

On the 20th September, an article appeared in the *Chronicle* saying that a second meeting had taken place in connection with the proposed institution on Thursday 18th September, so I suspect that the first one was held on the previous Thursday, the 11th. It was a meeting of the potential promoters, and it was reported that there were already 170 people who had indicated their willingness to support the idea, and it was expected that a General Meeting was to be arranged imminently.

Another preliminary meeting took place on the 23rd September, by which time the number of those interested in an involvement had risen to over two hundred. A provisional chairman was elected in the form of Mr. John Gilbertson,

Surgeon, and he called a General Meeting, to be held in the Corn Exchange on Lune Street on Tuesday 7th October.

It was immensely encouraging that despite the terrible weather that occurred on the appointed evening, an extremely large crowd, represented by people from all walks of Preston's life, assembled in one of the rooms in the Exchange. At the appointed hour of 7.30p.m Mr. Gilbertson was called to the chair, where he opened the business by explaining with great clarity, the nature and objectives of the proposed society.

I think it is indicative of the leading nature of the role Moses Holden played in the creation of the society, when he was the first person who spoke after the opening address and particularly with what he said to the gathering. He said, *"My fellow townsmen, it is with no small degree of pleasure I rise to move that a society now be formed, and that it be called "The Preston Institution for the Diffusion of Knowledge."*

He continued, *"Sir, I am impressed with the name, because it gives almost unlimited latitude to its members to acquire knowledge. Had such an institution be formed in the early part of my life, I would have thought it one of the brightest days of my existence. But, Sir, my opportunities were few; and it may not be generally known to my townsmen, that I had no regular instructions at school after the fifth year of my age, excepting a little at the Sunday school. The fact is, Sir, I had to make my way solitarily, and take as it were the pickaxe and cut my way through the solid rocks, without assistance or help from anyone; but I laboured at it; yes, I persevered and fainted not. While others were sleeping in their beds I was acquiring knowledge. Besides, the works on science were costly; I had those to procure alone, and did procure them with much labour and great expense. Whereas, in an institution of this kind, by uniting together, costly works are procured at a cheap and easy rate. General knowledge will be diffused to enlarge the mind; and when this is accomplished, it is lifted above the puny trifles of the day and has enjoyment in itself, nor can it be easily duped by sophistry, bigotry, or error, for it is the pursuit of truth, and these fall before that."*

He continued, *"This institution will have in it a library for the diffusion of knowledge, the youth by this means may acquire the history of his own and other nations of the world, and by perseverance store up the transactions of past ages as treasures for his own improvement. Natural history may be acquired at a small expense through these means, although works on this subject are so expensive. What I mean by natural history, not only taking in quadrupeds, birds, fishes, reptiles, conchology, animalcule, and entomology, (which last has become very popular, both with ladies and gentlemen), but botany and geology. Geography, or the situations of all the nations, seas, islands, mountains, rivers and lakes of the whole earth, may be acquired with pleasure.*

"Astronomy, which of all the other sciences enlarges the mind the most, together with geometry, may be fostered in such an institution. Biography for entertainment and good solid instruction; what can surpass it? Also voyages and travels, so that there will be something to suit everyone.

"But Sir, if we are pleased and satisfied with a smattering, or a skimming over the surface of the whole of these, instead of acquiring solid knowledge, and as it can be better effected collectively, than by individual effort, I sit down by moving that a Society be now formed, and that it be called the Preston Institution for the Diffusion of Knowledge."

The motion was seconded by Edward Makin, a cotton manufacturer, and carried into existence by the unanimous decision of the people assembled in the Exchange Rooms. There then followed speeches by a number of leading businessmen and dignitaries regarding the functioning of the Institution and the officers who were to run it.

Mr. Adam Booth, a twenty-seven year old mechanic then spoke of his proposal for the role of President of the Institution, saying that his nominee was both a resident in Preston and somebody who would meet with the widest approval. That man was the recorder of Preston, Thomas Batty Addison, and the mention of his name was met with a spontaneous round of applause. Another twenty-seven year old man, who was to become a leading barrister, Robert Segar, entered the discussion by observing that he *'was uncommonly pleased with the choice they were going to make of a president. I have no doubt that under the auspices of such a gentleman, the Institution would flourish'.*

Thomas Batty Addison then stood and paid wholesome tribute to the interest that had been shown in the formation of the Institute, and to the talent he had witnessed. He said that he was convinced that there was not only the will, but the ability among those that had declared an interest in the running of it, that it would become a most useful institution.

He concluded his speech by saying that in his opinion, such institutions were the means by which people could become more useful in their various situations in life and more intelligent members of society. He then added a word of caution, that nobody should expect such geniuses as James Watt, the Scottish inventor and engineer, to spring up as an everyday occurrence.

Finally, he added that whilst he had hoped that a person with more leisure time, and better ability to serve them had been chosen president, he would willingly lend his assistance, gave them his best wishes, and accepted the office.

In a speech that illustrated the link between the Temperance Movement and the Institute, John Johnson, a tailor, said that he took a very interesting view of the good effects that were likely to be derived from the Institution. *'One of the principle advantages would be the check that would be given to the indulging in that lamentable propensity, that many young men* were liable to fall into for want of better employment, of sotting away their leisure hours at the tavern, in drunkenness and gambling'.

I have included in Appendix Five, a list of those people who constituted the first committee of the Institution, and a list of those people who were memorialised on a stone in the UCLan Library, Adelphi Street, Preston, as being the Founders of the Institution. There are some who appear on both, and some who were involved on the committee but who were not regarded as being a founder member. There is a note at the bottom of the memorial stone to say that their Founder's Day is the 7th October.

A letter from somebody who was present at the meeting on the 7th October, and who signed him or herself as R.A. observed that several speeches were made, some of which were well conceived and fluently delivered. *'Among the speakers was Moses Holden, who gave the meeting ample proof of the extent and generality of his acquirements. Mr. Holden is obviously a self-taught genius; an enthusiast in science; his whole being seems wrapt in philosophical inspiration – there was an earnestness almost clothed in solemnity, in his style of speaking, that possibly, by some, might appear too serious for the subject; but there was a touching simplicity in his description of the difficulties he himself had encountered, unaided and unassisted, that obviously awakened the sympathy of the hearers. Mr. Holden is a man well calculated to direct the studies of the youthful mechanic, and impart to the society's portion of his own zeal and enthusiasm – without which it can never flourish.'*

R.A. went on to comment about the importance of the general interest of the gentry with regards to the Institute. The immediate need for donations of both cash and gifts of textbooks and instruments was considerable if it were to make a start that would give it the best possible chance of long-term success. The Committee had been charged with the responsibility of soliciting such donations, and they were successful in doing so.

Barely ten days after the inception of the Institute the number of subscribers stood at 382, with many of them, in addition to their annual membership fee of six shillings and sixpence (32½p), making cash gifts of five guineas, or equipment to a similar value. John Addison, a member of the Winckley Square set, and former mayor of Preston, donating twenty pounds worth of equipment. This equipment, it would seem, had been the property of the 1810 Literary and Philosophical Society, for which there are no records after about 1820, but by 1828 had not been officially dissolved. Another subscriber donated a thirty-six volume Encyclopaedia Britannica. By the end of October 1828, subscriptions stood at 485, and donations had reached well over three hundred pounds.

Cash and other donations were to be sent to the office of Robert Ascroft, the secretary, whose solicitor's firm operated in New Cock Yard, although plans were well advanced to acquire rooms in which to operate on the east side of Cannon Street. It was also their intention to set up a museum and library in the same property. In the meantime, the Institute met in the Corn Exchange, the place that saw its conception and birth.

When the Institute took its first steps, at the first official meeting, which took place on the evening of Thursday 13th November in the Corn Exchange, between six and seven hundred people were gathered. The number included twenty to thirty ladies. The president, Thomas Batty Addison opened the meeting by updating the audience on the current state of finances and membership. Barely a month into the venture, the number of subscribers had risen to 568, and their library could already boast 668 volumes, with the realistic prospect of it reaching 1,000 soon.

He reported that the Cannon Street rooms would be opened on Wednesday 19th November, complete with its library, and members would be able to begin borrowing books, limited to one at a time, on that date. Scientific instruments and items intended to form the basis of a museum were still being donated, and he reported that there would be no immediate expenditure made by the Institution on that type of thing to avoid unnecessary duplication.

In addition to that, he had to report that offers had been made to give lectures to members, as an alternative way of imparting knowledge. Mr. David George Goyder, a Glaswegian dissenting minister, who was involved in infant education, but also gave lectures on phrenology made such an offer, as did Moses Holden. Phrenology was a subject in vogue at the time, and another man, Dr. Alderson, of Hull, one of the most eminent physicians in England, also indicated a willingness to lecture in Preston.

On the 4th December 1828, the first lecture was given at the Institution, by Moses Holden. It was the first of a series on Optics. By this date, membership had risen to 595, and donations had climbed to five hundred and thirteen pounds, but by the time the first newspaper of 1829 was published it seems that the Institute was becoming a victim of its own success. A contributor calling himself 'A. Member' wrote to the editor of the *Preston Chronicle* saying that membership had risen to 670; the library had reached the 1,000 volume level and was expected to double during the ensuing year, but that the lecture room, whilst being a large room, could hold barely more than half the membership. The decision was made, therefore, that for Moses Holden's lecture on Optics in early December 1828, admission would be allowed to those members whose membership number was 300 or less, and the lecture would be repeated the following evening for those above 300. The lecture room was packed on both evenings.

A larger facility would have been useful for such popular lectures, but the Corporation had adopted a stance that prevented the Institute meeting in the Corn Exchange for 'the diffusion of knowledge', whilst they would still have allowed its use for meetings to organise such a society. Lighting, or rather the lack of it, was an additional issue in the Cannon Street rooms, but it was hoped that the relatively new Preston Gas Company, who had recently been generous in their grant of gas to the Shepherd Street library and to a fancy ball at the Corn Exchange, would *'cast a favourable eye on the Preston Institute'*. Already, the demand on books in the Institute's library was ten-fold the demand at the Shepherd Street facility.

Whilst the Corporation may have been reluctant to allow use of the Corn Exchange, the Earl of Derby was quite content to allow the Institute the use of the Cockpit behind the Minster. Apart from the instance of Moses Holden giving his Optics lecture there in April 1829; it had been used previously on the 5th March to present a lecture by Mr. Barton of Blackburn on 'Electricity'. There was a numerous audience, and the premises were deemed ideal for the purpose. The lecture included numerous interesting, and generally successful experiments. The less successful ones were not the subject of further comment!

Whilst the Institute grew and flourished as it enjoyed lectures on a wide and eclectic mix of subjects, I think that one more of the earliest of them is worth recounting. It took place on the 23rd April 1829, and was given by John Dewhurst, a mason and slater, and member of the Institute, who spoke about the history of architecture. The lecture took the form of a celebration of the perfection of the architecture of the Greeks and Romans, followed by a lament based on the current apparent decline in interest being taken in the subject. He concluded his lecture by encouraging his fellow members who were masons, joiners and smiths, to form themselves into classes, for the study of the qualities of stone, the strength of iron and timber, attributing the many accidents resulting from the collapse of buildings to a lack of this knowledge.

It was clearly his intention to give a series of lectures on this subject, for towards the end of this one, he announced that his next one would examine the state and progress of architecture since the invasion of the Romans to the present time.

He then presented the whole of the drawings, together with a beautiful model of the Choragic Monument of Lysicrates, which was known to modern Athenians as the 'Lanthorn of Demosthenes,' to the Institute. The model had been made by Thomas Duckett, a Preston sculptor, and is constructed in three distinct pieces; firstly, a quadrangular base; secondly, a circular colonnade, of which the intercolumniations were entirely closed; and thirdly, a cupola, with an ornament on the crown of the dome.

A typical example of a Lanthern of Demosthenes.

At the end of its second full year, membership of the Institute appears to have reached a plateau, with 551 people having paid their subscription. Funds were healthy, with around one hundred pounds in hand, and the library had grown to 1,700 volumes, including around 200, many of a valuable nature, which had been added in the last twelve months. The library was growing in activity, with the weekly circulation of books totalling three hundred. Several magazine subscriptions had been taken out by the Institute for use in the Reading Room, including the *Mirror*, the *Kaleidoscope*, the *Library of Useful Knowledge*, the *Natural History Magazine*, the *Quarterly Journal of Science*, the *New Monthly Magazine*, and the *Mechanics' Magazine*.

Although the Institute and its members have a vastly interesting history of well over fifty years, at the end of 1882, a meeting was called to wind-up the affairs of the Institute, and under the rules laid down in the constitution, the properties and other assets were handed over to the Trustees of the late Edmund Robert Harris, under the provisions of his will, that being the desire to 'establish and endow a Literary and Scientific Institution or other Charitable Institution or other Institution of Public Utility.' The new establishment became known as the Harris Institute, later the Harris Technical College, and ultimately the University of Central Lancashire or UCLan.

***

Having intimated that he would resume his touring and his lecturing, he began 1829 in the West Yorkshire village of Holbeck, when the *Leeds Mercury* of 24th January reported that he gave sermons in the afternoon and evening of Sunday 18th January, at the recently built New Chapel of the Wesleyan Methodists. One has to suspect that he has reached this place on the Leeds to Liverpool canal, being a village that by 1834 had earned the dubious distinction of being *'the most crowded, most filthy and unhealthy village in the country.'*

However, his travels seem not to have lasted long, for Moses returned to Preston where he gave his optics talk in the Cockpit, and by the beginning of July that year he has found it necessary to put an advertisement in the *Preston Chronicle*. He is soliciting teaching roles in the subject of astronomy, geography and the use of globes. The advertisement was addressed to 'the Heads of Families and Schools,' and notified them that his health will not permit him to travel during the summer of 1829. He noted that he was prepared to attend pupils either in their own homes or at his home at 7, Jordan Street. This was the only intimation that he had a health problem, and there is nothing to suggest what form of ailment it was.

Whatever it was, it didn't prevent him exhibiting his solar microscope in the Cannon Street rooms, where he had an instrument fixed into one of the windows for three consecutive days. The building that they used was on the eastern side of the street, so it is probable that the window referred to was south-west facing, and overlooked Cannon Street. Once again, it was the 'wonderful and striking facts of nature' that the instrument was capable of displaying to their best effect that proved the most popular.

On the 5th November 1829, Moses repeated his third Cockpit Optics lecture, this time holding it at the Theatre Royal, where nearly one thousand people were present. This episode of the series was the one where he demonstrated some optical illusions which were of an amusing character, and using the various lenses to expand or diminish the size of objects painted on a transparent material, usually glass. This was typical of the entertainment that was described as 'Phantasmagoria'.

Correspondence with the newspapers during this decade had been somewhat sparse, but whether that had been because there had been little to record that would have benefitted the amateur isn't known, but once again in 1830 there were two letters, three weeks apart, to the editor of the *Preston Chronicle* relating to spots on the Sun, or Sunspots. He included descriptions of their movements across the surface of the Sun, estimates of when they would disappear behind it, and estimates of their size.

It is interesting to note that Moses measured the Sun's diameter at 836,149 miles, and the largest of the spots at 49,230 miles. With the benefit of modern technology, diameters of the Sun still vary from 864,938 to 865374 miles, which means that his estimates were quite a credit to his ability at that time.

Incidentally, Sunspots are temporary phenomena on the photosphere or outer shell of the Sun, which appear as dark spots by comparison to the area around them. They are caused by extremely intense magnetic activity. They usually appear as pairs, with each of the pair having opposite magnetic poles to the other.

His final series of lectures of this decade were held at the Theatre Royal in Preston on the 26th, 27th and 29th April, when again, his subject was astronomy. Editorial material in the *Preston Chronicle* prior to the event suggested that there would be new material included compared with his last lectures here, but later reports suggested that although the Boxes in the Theatre were packed, the other parts of the house were not so full. A less than full house was something new for Moses, but it was reported that on the whole, the course had yielded a return which has fully satisfied the hopes, and more than fulfilled the expectations of the lecturer.

We have seen before that Moses held events at his home to demonstrate his solar microscope, and he held a further such event in the middle of June 1830. It was reported that he had fitted up an apartment in his house in Jordan Street, *'for the purpose of exhibiting to his friends, the wonders displayed'* through the means of this instrument. It went on to say that those who had never witnessed anything of this kind, could not fail to be filled with admiration and amazement, on seeing the beauty displayed by minute particles of organised matter, which are too small to be either seen or felt by the unaided senses.

There has never been a clue as to who constituted 'his friends', but based on my knowledge of the sort of people he generally associated with, it will have been representatives of the manufacturing classes, the professionals from Fishergate and Winckley Square, and other local dignitaries. In other words, they were probably those persons with influence who were able to offer their support for his efforts.

In chapter seven we'll discover ways in which Moses tried to promote himself, not just for the purpose of his lecture events, but also his Celestial Atlas and his Almanac.

Chapter Seven

1831 – 1840

The opening year of this decade will see Moses Holden reach his fifty-fourth birthday, but there is no hint that he is thinking of slowing down. In fact, this period will see some momentous occasions, such as his involvement with the early days of the Temperance Movement, his election as a Freeman of the Borough of Preston, and a continuance of his tour of lectures.

The first half of 1831 saw Moses cruising along the Lancaster Canal to deliver his course of lectures in the theatre in Lancaster at the end of April, and then exactly a month later he delivered the same course in the Theatre at Kendal.

In an article in the *Lancaster Gazette* prior to his engagement in that town, it made the point that it had been several years since his last visit there. It was one of the first places he lectured in 1815, but whether he'd been back since that date isn't known, but the article mentioned that 'opportunities like the present one seldom occur in Lancaster.' The people of Lancaster had formed a Mechanic's Institute in 1824, but it hadn't been as successful as the one in Preston. In fact the newspaper recorded, after mentioning the success of the Preston Institute, 'that they were happy to find that our humbler establishment was about to avail itself of an opportunity like the present.' For this event the Committee of their Institute were making arrangements for their members to attend, as a body, all three lectures.

I made reference earlier to the Institute for the Diffusion of Knowledge, and the Temperance Movement being in many ways complementary to one another, with many of the personnel common to both interests. There are those with a close interest in the subject of the early days of temperance activity, who deny that Moses Holden had any part in it, but there is credible evidence to show that he had.

Taking a step back for a moment, a temperance society was formed on the 1st January 1832; the young men attending Joseph Livesey's Sunday school formed themselves into such a society. Though one of the individuals involved, John Broadbelt, advocated total abstinence at the time, he was over-ruled. It's strange how ideas can be over-ruled, only to be reintroduced by one of the objectors at a later date; I'm sure there must be a word for it, and if there isn't, there ought to be! However, this society very shortly merged with the Preston Temperance Society that was formed, as we will discover, in the next paragraph.

In "The Middlemost and the Milltown's: Bourgeois Culture and Politics in Early Industrial England," Brian Lewis wrote, *'At a meeting in the Theatre on March 1832, chaired by Moses Holden, astronomer and Methodist preacher, a deputation from the Blackburn (Temperance) society – the Rev. Francis Skinner and George Edmondson, Quaker schoolmaster of Lower Bank Academy – proposed the formation of a Temperance Society in Preston. The proposal was seconded by the Rev. Richard Slate, an Independent minister, and all the local clergy and ministers were declared ex officio members of the committee of the Preston Temperance Society.'* At the end of the meeting, ninety people joined the Society, and very quickly that grew to two hundred.

On the 11th July 1832, a Temperance Tea Party was held at the Corn Exchange under the auspices of Moses Holden and Joseph Livesey and other active members of the Society. Between five and six hundred people were in attendance in the Cloth-hall of the Exchange, the greater part of who were females. Tea, coffee, bread, buns and muffins were greatly enjoyed, with the tea being offered as 'the beverage that cheers but not inebriates.'

Soon after seven o'clock, Moses Holden was called to the chair, where he delivered an address at great length on the evils and dangers of inebriation, and the advantages and blessings of abstinence from intoxicants. Membership of the Society now exceeded one thousand, and whilst congratulating those thousand, he said that he felt it likely that that figure would be ten times as big within twelve months. The following day, a field meeting of the same society was held on Preston Moor – now known as Moor Park. Speeches were made from the back of a horse-drawn wagon, decorated with an appropriate flag, to an ever changing group of people over a four hour period, and for whom wooden forms had been provided. It was estimated that no fewer than two thousand people attended, each of whom were said to have listened with the greatest attention and respect.

In places other than Preston, the Temperance Movements had been non-sectarian and non-political organisations, but it quickly became clear that the Preston movement was being politicised by the likes of Joseph Livesey and other Non-conformist, shop-keeping radicals, who Livesey represented through his journal, the *Moral Reformer*. Moses Holden was a man of tolerance and moderation, but the circle that surrounded Livesey believed that the drinking of beer was as bad as that of gin, and whilst the temperance men were travelling between manufacturing areas spreading the word, Livesey and his followers were churning out written tracts like the *Preston*

Temperance Advocate from 1834, promoting total abstention from alcohol.

At its outset, the Preston Temperance Society attracted large numbers of people of influence in the town, but once total abstinence was voted for in 1835, cotton manufacturers, surgeons, and local Members of Parliament, their interest and support dwindled. It is said that Moses' interest also drained, not because he consumed alcohol, which he didn't, but because those that were withdrawing their support for the movement were the ones from whom Moses depended on for his lecturing and instrument making interests.

Total abstinence was even opposed by the Rev. John Clay, the chaplain of the towns' prison, and zealous moral reformer; but whilst a supporter of the temperance society, was another who refused to convert to teetotalism saying, *"To deprive the people of their chief animal gratification, while they are still incapable of any gratification which is not animal, would be a dangerous experiment."* However, his influential Annual Reports, hammered away at the need to tackle drunkenness, and the hardships and poverty it tended to produce.

It was around the time that Moses severed his links with the temperance movement that he began seeking accolades and probably soliciting letters of reference from those he had done work for. Possibly the first such letter was one from George Horrocks, the cotton manufacturer, in November 1832. It took the form of a commendation for Moses' work in making two convex lenses, extolling the quality of them, and comparing them with ones he had previously bought from Dolland of London. We have already seen that they were leading manufacturers at the time. With Moses' lenses the clarity and definition of closely grouped stars was enhanced by distinctly separating them from one another, and showing them to be perfectly round; qualities that aren't possible with an inferior lens.

George Horrocks concluded his letter by saying, "I suppose you esteem Dolland equal if not superior to any in London for making these things well, and it is only with them that I am able to make a comparison, and as far as I am able to judge of them, those which you have made for me, are quite as good, if not better than any I have in my possession."

Although there is nothing in writing, the Vicar of Preston, the Rev. Roger Carus Wilson, himself a capable astronomer, once offered the opinion that a telescope that Moses had made for him was better than one he already possessed that had been made by Dolland. A similar comment, naming the same manufacturer, was also made by Sir Hesketh Fleetwood. I suspect that the number of his clients, who made favourable comparisons with the leading manufacturer of the day, was a response to a carefully worded rhetorical question.

We will shortly learn about letters that Moses wrote to royalty, through the offices of local Members of Parliament, but before we do, I'd like to refer you to a series of lectures that Moses gave in his adopted Preston in April and his native Bolton in October 1833.

His April engagement was his long-standing triennial series, for which no particular preparations were evident, but following their completion a report in the *Preston Chronicle* claimed that on each evening the boxes were crowded to excess, particularly on the first night, when *'several parties who were unable to find convenient seats there, took their stations in the pit.'* Both of Preston's Members of Parliament sent in their subscriptions for the three-night course, but were unable to attend; and in consideration of this, Moses voluntarily admitted forty of the older boys from the National school to attend the lectures.

I can only imagine that the spring and summer of 1833 were, unlike the Preston lectures, full of preparation. Firstly, in addition to the usual advertising handbills that were prepared for his lectures in the places he was to visit, he produced a further handbill which was intended to introduce him to potential audiences. It was a one-sided document, the top half of which contained the names of twenty-six of Preston's leading dignitaries, ranging from the Vicar of Preston, the Mayor and the Recorder, to the Chaplain of Preston Prison, the Town Clerk, and former mayors like Nicholas Grimshaw. A note at the foot of the flyer invited readers to inspect the original document, with the signatures of the named gentlemen, at his home in Jordan Street.

The lower half of the flyer was an extract of a letter from his friend, William Rogerson of the Greenwich Observatory. Rogerson was an East Yorkshire man from Pocklington, and in his letter he refers to having attended a lecture of Moses' in Hull. It is probable that this was how they became acquainted, and several letters from him still exist.

TESTIMONIALS

OF

MR. HOLDEN'S MERIT.

Preston, September 12th, 1835.

We, the undersigned, have known Mr. MOSES HOLDEN for many years, and have attended his Lectures on Astronomy, in every department of which Science he is well versed. The Lectures contain a variety of interesting matter, illustrated by Machinery of a peculiar and ingenious construction, invented and principally made by himself. Independently of the knowledge and skill which Mr. Holden has acquired by extraordinary perseverance, under circumstances of great discouragement in early life, his general character in this Town, where his abode has been fixed from infancy, fully justifies us in recommending him as deserving of support from the respectable Inhabitants of those places in which he is about to deliver Lectures :—

ROGER CARUS WILSON, M.A., Vicar
  of Preston.
JOHN ADDISON, Mayor of Preston.
T. B. ADDISON, Recorder.
R. W. HOPKINS.
CHAS. BUCK.
B. F. ALLEN.
DANIEL NEWHAM.
GEO. HORROCKS.
R. BROWN.
RICHD. PALMER, Town Clerk.
JOHN CLAY, Chaplain to the House
  of Correction, Preston.
THOS. CLARK, Curate.
THOS. LEACH.

SAMUEL CRANE.
J. RIGG, Minister of St. Paul's.
JAMES PEDDER, Banker.
JOHN SWAINSON.
WM. TAYLOR.
W. ALEXANDER, M.D.
J. H. NORRIS, M.D.
A. MOORE, M.D., Oct. 1835.
WM. ST. CLARE, M.D.
JNO. LAWE.
N. GRIMSHAW, Senior Alderman
  of Preston, 1835.
S. HORROCKS, Jun.
GEORGE JACSON, Oct. 1835.

EXTRACT OF A LETTER

FROM THE ROYAL OBSERVATORY, GREENWICH,

MAY 1st, 1833.

"When you exhibited your Transparent Orrery, I was indeed highly delighted: the appearances of the Moon and Planets were so natural, that I imagined myself looking through a powerful Telescope; also the appearance of solvible and insolvible Nebulæ through Optic Tubes, &c. In fact, the whole of your grand exhibitions, with the simple and clear instructions you conveyed by your lectures, left upon my mind sensations which I can never forget; and though I saw your scenery and heard your descriptions repeated at Hull, and in other parts of the kingdom where you kindly invited me to visit you, I was never weary of their repetition; and would deem it an indescribable pleasure if I were so situated, to visit Preston at the present time, and witness the whole once more."—"Yes, my dear friend, although I am every day in the first Observatory in the world, and using instruments of the very best kind, and viewing by them the belts of Jupiter, the ring of Saturn, the frozen poles of Mars, the horns of Venus, the spots of the Sun, and the Moon's varied disk, 'rivers and mountains on her spotted globe;' yet, my friend, I should deem it a mental feast, to see and hear your praiseworthy performances."

WILLIAM ROGERSON.

*** The original MS., with the Signatures of the above Gentlemen, may be seen at the Lecturer's.

WILMER AND SMITH, PRINTERS, CHURCH-STREET, LIVERPOOL.

The flyer was dated the 12th September 1833, and it heralded a tour that had been reported in the *Preston Chronicle* of the 7th of that month. It revealed that the first course would be at Chorley, followed by Wigan, Bolton, Manchester, Oldham, Stockport, &c. in succession. It is interesting that the list appears incomplete, with others possibly following on from Stockport?

A lengthy article appeared in the *Bolton Chronicle* of 26th October that year, and yet there was no mention that Moses was a native of that town. It was, however, very descriptive in many ways, and certainly worthy of spending a minute or two discussing. As was often the case, Moses had given a recapitulatory lecture following his usual course; but, unfortunately the night had been excessively wet, and had been disappointing from an attendance point of view.

The article reported that, *"He exhibited nearly the whole of his splendid views, fifty in number, illustrating the appearances and revolutions of the heavenly bodies with the minutest exactness. Mr. H. Is a striking instance of what may be accomplished by perseverance and industry, being a self-taught man, and having arrived almost at perfection in his studies of this sublime science."*

There followed a number of comments that came as a surprise, but they were, probably just one man's opinion. He said, *"The only drawback on Mr. Holden's qualifications as a lecturer is, that he is no orator, and occasionally provincialisms intrude into his quotations of the most beautiful passages of our celebrated authors. This defect grates harshly on the refined ear, but is amply counterbalanced by the sincerity of intention manifested by the lecturer, so to simplify his observations that they may be comprehended by the meanest capacity."* What a pity he didn't give some examples!

The writer then made comment about how Moses never let an opportunity pass him by without trying to make a connection between the heavens and with God, something that I believe

Moses did out of habits of a lifetime. There was no doubt that he regarded the mysteries of the universe as the work of God.

His observations on the usually difficult to understand theory of the tides was next commented on, and illustrated by a view in which the influence of the earth and moon upon them was exhibited, and the writer recorded that he had certainly succeeded in placing the subject *"in a much clearer light than any other lecturer we have ever heard."* Moses then said, *"It has cost me no less than 4,000 observations to arrive at the knowledge that I possess, of the causes of the ebbing and flowing of the sea, and I have no doubt that I can acquire much more information by the same means."*

Finally, he spoke about *'those eccentric bodies'*, Comets. He explained that there was to be a Comet visible in 1835, but that the closest it would ever be to the earth was eighty-nine million miles distant. He then added, somewhat sardonically, *"So there will be no occasion to indulge in those ridiculous and absurd fears generated by German astronomers that this body would pass so near to the orbit of our earth as to cause its dissolution!"*

An article in the *Manchester Guardian* of 11th January 1834 opened by saying that Mr. Holden, whose splendid appearance gave so much satisfaction a few weeks ago in a course of lectures at the Theatre Royal, is to deliver two more courses of lectures during the following week, commencing 13th January, this time in the Queen's Theatre. It made the observation that they did not know of any exhibition which, to young people especially, combined so much amusement and instruction as a lecture on astronomy, illustrated by good apparatus; and certainly that used by Mr. Holden is beyond all comparison the best we have ever seen.

Manchester has had three Theatre Royals. Moses lectures will have been at the second one that existed in Fountain Street between 1807 and the time it was destroyed by fire in March 1844.

August of 1834 saw the introduction of Moses Holden's Almanac, so to avoid any confusion I would like, first of all to mention that from 1770, the Rev. George Holden, one time Curate of the chapel at Pilling, and his brother Richard, produced the 'Holden Almanac and Tide Tables'. It went on to be produced for a further one hundred years by members of the same family, and was then produced by other owners for a further century, with the 205th and final edition being in 1974. I mentioned in the early part of Chapter Three that there was no relationship between Moses and this family, unless of course, it goes back further than records are available.

Church Street, Preston,
August 20th, 1834.

Mr. ADDISON

HAS the honour to announce to the Merchants, Bankers, and Tradesmen of Preston, that he has made arrangements with

MR. MOSES HOLDEN,

ASTRONOMER AND LECTURER ON THAT SCIENCE,

To Publish Annually

A BOOK & SHEET ALMANAC,

Both of which will appear in November next, for the year 1835, being the third after Bissextile, or Leap Year, and will be entitled

"THE

PRESTON ALMANAC

AND CALENDAR

For the Northern Division of the County Palatine of Lancaster,'

BY MOSES HOLDEN,

ASTRONOMER.

The Calendar will be accurately calculated by Mr HOLDEN, and will shew the

TIMES OF HIGH WATER AT LYTHAM,

And the Heights of the Tides every Day in the Year;

THE RISING AND SETTING OF THE SUN AND MOON EACH DAY, CALCULATED FOR THE MERIDIAN OF PRESTON;

In the Calendar of the Sheet Almanac, will be carefully marked for the convenience of the Attorneys generally of this County, the Days of Holding the Courts of Quarter Sessions at Lancaster, Preston, Salford, and Kirkdale; the Return Days, of the several Courts of the County Palatine of Lancaster; and all such other Public Days, as may be generally useful to the Profession of the Law. The Holidays of the Church and all other Public Days will also be carefully marked.

The Miscellaneous Register of Information
Will contain several
ENGRAVED VIEWS OF CONSTELLATIONS,
Drawn by Mr. HOLDEN,
AND A VIEW OF THE COMET,
As it will be seen in October, 1835, with
ASTRONOMICAL FACTS & PHENOMENA,
and explanations of
THE CELESTIAL CHANGES OF THE YEAR.
And a Variety of Useful Information

TO FARMERS AND COUNTRY PEOPLE
Relating to Crops, and the most seasonable time for cultivating and husbanding the same.

A Correct List of the Fairs held in the County, and of useful Stamps, Taxes, &c., will be carefully selected. And the following among other Local Information, added:—

A List of the Corporation of the Borough of Preston—The Acting Magistrates for the Hundreds of Amounderness, Blackburn, Leyland, and Lonsdale—The Public Offices and Officers in the County, resident in Preston—A List of the Bankers for all the Principal Towns in the County, and upon whom they Draw in London—A List of Insurance Offices, with their Agents in Preston—The Post Office Regulations—And all such other Information as may be considered Useful.

Moses' Almanac was commissioned and printed by Addison's booksellers and printers, and produced for the Northern Division of the County Palatine of Lancaster. The Almanac was for the year 1835, and gave daily times and heights of each tide at Lytham. It gave rising and setting times for both the Sun and the Moon, calculated for the meridian of Preston. In addition to dates for the various Courts in the County, together with other information 'useful to the Attorneys', it also contained engraved views of the constellations, drawn by Moses, and a pictorial view of the Comet that would be visible in the August, September and October of 1835, 'with astronomical facts and phenomena, and explanations of the celestial changes of the year'. This Comet was the one known as Halley's Comet. There followed a section for the farmers and other country people, 'relating to crops, and the most seasonable time for cultivating and husbanding the same'.

Much of the astronomical information that he used in the Almanac was received from Greenwich Observatory and his friend William Rogerson, but a lot of local information, including a list and dates of the Fairs to be held in the County, together with information about the town's officials, the Corporation and principal public officers.

In the April prior to the publication of the Almanac, Moses had been conferred with the honour of the Freedom of the Borough. At a Council meeting on Friday 25th of that month, he was appointed a burgess without solicitation, to mark the sense entertained of the great merits of our respected townsman. A month later, on the 30th May 1834, the official presentation of the honour was made, in open council, on which occasion, the Recorder, Thomas Batty Addison passed a high and well-merited eulogium upon the new burgess; Moses replied 'in a feeling and grateful manner.'

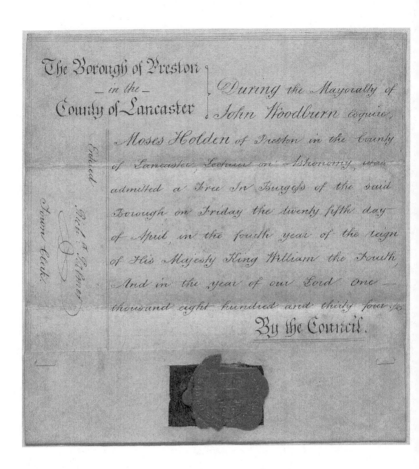

The Borough of Preston
— in the —
County of Lancaster

During the Mayoralty of
John Woodburn Esquire,

Moses Holden of Preston in the County of Lancaster, Lecturer on Astronomy, was admitted a Free In Burgess of the said Borough on Friday the twenty fifth day of April, in the fourth year of the reign of His Majesty King William the Fourth, And in the year of our Lord one thousand eight hundred and thirty four.

By the Council.

Certificate of Freedom of the Borough

It would seem that his elevation to a position few people ever reach seemed to give him fresh energy in terms of self-promotion. On at least two occasions he wrote to members of the royal family to seek approbation for both his Almanac and his Celestial Atlas. The first reply he received was dated the 16th July 1834, and from St. James' Palace. It read:

'Sir Henry Wheatley presents his compliments to Mr. Holden, and has the honour to inform him, that the small Celestial Atlas, and the copy of the Preston Almanac, which Mr. Holden has forwarded for the King's approbation and acceptance, have been submitted to His Majesty, and His Majesty has been graciously pleased to receive them'.

Sir Henry Wheatley was the Keeper of the Privy Purse for both King William the Fourth and Queen Victoria from 1830 until 1846.

In 1835, Moses received a reply to a letter he had sent through M.P. Mr. Hesketh Fleetwood, dated 4th April, from that same man. He also enclosed a letter from Sir John Ponsonby Conroy, a British Army officer who served as comptroller to the Duchess of Kent and her young daughter, Princess Victoria, the future Queen. The letter from Sir John read,

'Sir John Conroy presents his compliments to Mr Hesketh Fleetwood. He has had the honour to lay his note, with Mr. Holden's letter and Almanac before the Duchess of Kent, and Her Royal Highness will feel much obliged to Mr. Hesketh Fleetwood, if he will inform Mr. Holden that the Princess Victoria accepted with much pleasure and interest, the "Almanac"'

Kensington Palace

7th April 1835

At the end of April 1835 we get a rare insight into Moses' early life and into his thirties. When William Taylor, an employee of Horrocks & Co., cotton manufacturers, and probably involved on the technical and foundry-work side of the business for that company, retired, Moses was an invitee. At one point in the retirement celebrations Moses addressed the gathered assembly and said, *'That it had been a gratifying occasion when he first met Mr. Taylor. I have grateful feelings, and I will have throughout life, towards him; for it was he who had taken me by the hand when I had hardly anyone else to help me. It was well-known that if it had not been for him and the people belonging to the house of Horrocks & Co., I would hardly have been able to advance myself as I thank God I have done.*

'When I was preparing the orrery with which I have travelled through most of England, I received the greatest assistance from them; and indeed, many parts were made at the mechanic's shop belonging to the Moss Factory, by persons who had more experience in calculating the teeth of wheels than I possessed.

'Some people might suppose that my scientific pursuits might render me cold and insensible, but I can assure you that I have a warm heart, and a grateful recollection of the many kind acts which have been done for me by my friend Mr. Taylor, and they will animate my mind so long as sense and feeling is granted to me.'

In the September of 1835, the tables were turned, as it were, when Mr. Thomas Walker, an astronomer, came to Preston to deliver a series of lectures. They were held in the Assembly Room of the Bull and Royal Hotel. There were two lectures in the course, and the original intention was that he would deliver them twice, on the 7th and 8th of September, and then repeat them on the 10th and 11th.

He was advertising the use of a 'newly invented Grand Concave Orrery, 42 feet in circumference, which shows the arrangement and astonishing movements of the Celestial Orbs, more in accordance with nature than any orrery that has ever been hitherto seen in the country.' A transparent orrery, like the one Moses used required a theatre to be completely dark, whereas this equipment could be viewed in a lit room. He was also intending to give a demonstration of his experiments in ærostation, a term first applied in 1784 to the art or science of operating lighter-than-air aircraft, 'which will be illustrated by a small machine flying over the heads of the audience.'

The charges for the lectures were quite modest at two shillings (10p) for front seats, and one shilling (5p) further back. Youths under fourteen years of age would be allowed to occupy the front seats at half price. However, despite the modesty of the charges, the attendances were reported to be disappointing, but that fact wasn't mentioned in the editorial of the *Preston Chronicle* at the end of the first week. Their report was particularly positive.

In a separate column of the same newspaper were two letters from Moses Holden. One was a report on the progress of Halley's Comet that I referred to in connection with his Almanac, but the other one was in connection with Mr. Walker's lectures. He wrote:

'I have attended Mr. Thomas Walker's lecture on Astronomy this week, and found his orrery a very ingenious and curious machine, calculated to give a very good idea of the movements of the Solar system.

In this lecture he is very familiar with his subject, and makes everything very plain, so that a person of ordinary capacity may understand him.

I have written this because he has been poorly attended, and I thought it might be in consequence of the Preston people being taken by the two last lecturers on Astronomy who visited this town, and had nothing to exhibit worth seeing, nor did they understand anything of the science.

Mr. Walker understands his subject and those who may feel disposed to hear a lecture on Astronomy would do well to avail themselves of this opportunity.'

Clearly undisturbed by a competitor offering the same type of entertainment as himself, and despite the claim in the advertisement that Walker regarded his equipment as 'superior to anything else in the country', which would include that used by Moses, I find it typical of the man to be offering fatherly encouragement to a thirty year old lecturer who was closer to the spring rather than the late autumn of his career.

The lectures were offered as a further repeat on the Monday and Tuesday following Moses' letter, but there is no record as to whether it had any effect on attendance. There was, however, a follow-up letter from Moses in the *Preston Chronicle* of the 19th September, regarding Halley's Comet. In it he reported that, owing to the poor weather, the Comet had been difficult to observe, 'but with a new Comet-finder I have recently made, it was very distinguishable.' When the visibility improved later in the week, he reported that 'I tried several telescopes, as well as powers (strength of lens), to see how small an instrument it could be seen with, and I must say I saw it very faintly once or twice with a very good opera glass which magnifies only twice.' In a similar way that he weekly reported his tracking of a Comet in the *Preston Journal* in 1811, he did much the same thing in this instance in the *Preston Chronicle.*

The tail-end of 1835 saw two sets of lectures being delivered in Liverpool, surely the most frequently visited city by Moses. To offer those lectures in consecutive weeks, the one's commencing the 2nd and 9th of November, was highly unusual, but towards the end of the second course, a notice appeared in the *Liverpool Courier* which read, 'Moses Holden continues to give his lectures to fashionable and increasing audiences;' so clearly the demand was there. In fact, as we will discover in 1844, he gave nine performances in Liverpool in eleven days; astounding!

Christ Church, Preston

From the time of arrival of the first incumbent of Christ Church, the Rev. T. A. Clarke, Moses seems to have struck up a friendship with him, and he became a regular attendee. He was later appointed Verger of the church. They were the bells of this church that 'rang out, deeply muffled' on the night that Moses died in 1864.

The erection of a building that was to play a significant role throughout the rest of Moses' life was begun to be constructed at the northern end of Jordan Street in 1835; it also had an entrance from Bow Lane. That building was the Norman-style Christ Church, the only church in Preston to be built of limestone, and it stood just yards from Moses' house. Despite the fact that it was intended to cater for Church of England devotees, Moses was to become one of those followers, with a close friendship being forged with the incumbent the Rev. Thomas Clarke. It was consecrated towards the end of 1836, and opened its doors for the first time the year after that. Almost twenty years later, in 1853, Moses was elected as a churchwarden of Christ Church.

The 1830s was a period when a number of churches were being built, and when the foundations of St. Andrew's Church on Blackpool Road in Ashton had been laid down in August 1835, a time capsule, in the form of a glass vessel, was deposited in those foundations, containing memorials of the time, and one of those memorials was a copy of Moses' Almanac. It would surprise me not to learn that a similar capsule had been buried in the Christ Church foundations, but I have found no record to indicate such an event.

When it was first erected funds were insufficient to provide the church with a clock, and it was over twenty years before it was to get one. However, when it did, it was considered to be the clock that displayed the most accurate time in the whole of Preston. The reason for that was that Moses Holden would calculate the precise time from his astronomical clock in his observatory, and his observations of the Sun, with the necessary information being passed to the man charged with the upkeep of the clock. It was reported when it was first installed that 'it keeps time remarkably well, and can be depended on to a nicety,' and this arrangement continued for the following six years or so, until Moses died in 1864. We will see shortly, that

Anthony Hewitson suggested something other than the fact that it was Moses Holden who provided the precise time for the clock, for he had been named as the person responsible in a letter of commendation in Hewitson's rival newspaper, the *Preston Chronicle*, on the 27th February 1858.

Clockwork housing

When Christ Church was taken over by the County Council, the clock was electrified. The old clockwork housing is now displayed in one of the basement corridors of the County Offices. It was said that this clock was the most accurate in Preston, due solely to the fact that Moses Holden calculated the time from his observations of the Sun, and the findings communicated to the man charged with adjusting the clock.

When it was first installed, an article appeared in the *Preston Chronicle* detailing the maker and installer as William Brown, watch and clock maker of Fishergate. It was said to be replete with all the recent improvements in clock-making, and was said to be an exact model of the noted iron clock in St. Ann's Church in Manchester.

The dial of the clock was described as a skeleton one, of cast iron, and five feet eight inches in diameter. The hands are of a fleur-de-lys pattern, large and bold, and may be well seen at a distance; the pendulum is about thirteen and a half feet long.

The newspaper article that introduced the clock to their readership gave a sense that being made of cast iron, it would have its critics; and they were right. There followed a series of letters in the *Chronicle* that gave an opportunity for interested parties to voice their opinion. As the original article predicted, again accurately, clock-makers who had missed the opportunity to install a clock themselves, were the ones who seemed to have the loudest voices and strongest opinions. One such was Jonathan Simpson, clockmaker of Lune Street in Preston, who, after criticising the use of cast iron instead of brass in the construction of the clock, added, 'I cannot say anything more about it without seeing it, and I have been told that I cannot see it.'

In 'The Story of Christ Church', by Harry C. Pye (1945), it was said that the clock 'was provided mainly through the exertions of the then parish school-master, Thomas Jackson Bakewell, and for many years the clock was under his personal care, steadily gaining local fame as an excellent time-keeper.' Anthony Hewitson, writing under his pen-name *Atticus* in the *Preston Chronicle* wrote, 'It (the clock) is looked after by a gentleman learned in the deep mysteries of horology, who won't allow its fingers to get wrong one single second, who used to make his own solar calculations in his own observatory, on the other side of Jordan Street, who gets his time now from Greenwich. He is

142

thoroughly master of the clock, and could almost make it stop or go by simply shouting or putting up his finger at it; and who would have just cause to find fault with the sun if antagonising with its indications.'

I would take issue with *Atticus's* suggestion that the person that was the horologist was other than Moses Holden, and the only reason that Bakewell 'now gets his time from Greenwich', was because Moses Holden had died. Moses is given no credit in either of these accounts for his fund-raising efforts or his many endeavours to ensure the accuracy of the clock.

Bakewell was described in 1853 as 'Master of Christ Church National School,' and in the Census of 1861 he was said to be an unmarried fifty eight year-old retired cotton spinner, living at 17 Jordan Street, and as having two servants. The address ties in neatly with Anthony Hewitson's comments about his horologist living on the opposite side of Jordan Street. Moses lived at number 14 after the numbers began to run alternately, with a consequence that they will have lived directly opposite one another.

The front of Christ Church was retained and incorporated in a very sympathetic manner into a new building when the Lancashire County Council offices were extended in the mid-1970s, and now forms part of their premises. The original clock face is still in place, with traces of the original French-blue colouration still detectable in the outer recesses of it, but the original fleur-de-lys fingers have been replaced. The wonderful clock-work mechanism that ensured its reliability for 140 years has now been replaced by electricity, with the original proudly displayed in a basement corridor in the County Offices.

Moving forward a few years to 1844, but still long before the advent of Christ Church clock, Moses wrote to the *Preston Chronicle* regarding an approaching total Eclipse of the Moon. He gave all the relevant timings for the start and end of the event, adding a note that read, 'If any person wishes to observe the Eclipse at the correct time for Preston, I shall feel a pleasure in giving them the time, as the public clocks in Preston are not to be depended on.' Things had to be correct, and when Christ Church finally got its clock, we'll discover later what part Moses played in keeping it correct.

***

1836 was a Triennial Lecture year, and in mid-March an announcement was made giving several weeks' notice of the event; the lectures were delivered week commencing 18th April, with 'a most numerous attendance' being recorded afterwards. It expanded on that comment by adding, 'The Boxes and Pit have been filled to overflowing.' Records of the cash receipts from his lectures are extremely scarce, and the only ones available other than those for his Farewell Triennial Lectures in 1852, are for 1836 and 1839. They are not as detailed as those in 1852, but they do tell us that the total receipts amounted to one hundred and seven pounds and five shillings and six pence (£107.27½p), and the expenditure totalled nine pounds and eleven shillings (£9.55p). Again it's difficult to contextualise these figures, but in 1836 a mill-worker would earn no more than 75p per week.

A breakdown of the expenditure reveals that three pounds and three shillings (£3.15p) was the charge for the rent of the theatre – a guinea each day. Printing and advertising materials cost three pounds three shillings and sixpence (£3.17½p), with the rest being split between fourteen other areas of expenditure. What it didn't include was for any work in assistance of his lectures, for the orrery needed turning and winding to make it work. His son, William Archimedes would have been about twenty years of age at this point, and I believe that, as he did in later lectures, will have been helping his father in this instance. Perhaps, unlike 1852, he didn't get paid? In the case of the 1839 lectures, the receipts were seventy five pounds seventeen shillings and sixpence (£75.87½p), with the rent having risen to five guineas (£5.25p) for three nights, and with three people paid for helping (Clare 65p., Edward 22½p., and Ward 20p.), the expenditure was eleven pounds one shilling and sixpence (£11.07½p). A report following the course suggested that the first two evening's attendance had been seriously affected by unfavourable weather, and that the improved circumstances on the final night saw a highly respectable attendance, 'with every part of the house amply tenanted by attentive auditors who testified their gratitude by repeated marks of approbation.'

Shortly afterwards, on the 14th May, an article by Moses appeared in the *Preston Chronicle,* announcing the impending Eclipse of the Sun. He wrote, 'The Eclipse that will take place on Sunday 15th May, will be the largest obscuring of the Sun I will have ever seen, nor have we had one as large at Preston since the year 1764, which happened on a Sunday, the 1st April, in the middle of the forenoon. Divine service in many of the parish churches in Yorkshire was postponed, as several very aged persons can recollect. My friend, Mr. Rogerson of the Royal Observatory says, *"I have heard my grandfather say, that at Pocklington in Yorkshire, the gloom was such that the minister could not see to read the prayers, and the congregation stood in the churchyard gazing at the Eclipse."*

In an article that reported on the event, Moses said that Venus had also been distinctly seen at the time of the greatest obscuration and that those with a keen eye would also have observed Jupiter; but, owing to the uncommon and unexpected brightness of the day, the darkness during the eclipse was not as great as predicted. It was, however, sufficiently so to deceive many birds, which, believing that dusk was approaching, retired to their roosting places. Moses summed up the eclipse by comparing it to the appearance of an approaching storm.

The years 1837 and 1838 saw Moses arrive at and pass his sixtieth year, and one may be forgiven for thinking that he might be beginning to slow down. However, the available material tells a slightly different story.

In May of 1837 he made another visit to the theatre in Lancaster, where, on the 22nd, 23rd, and 24th he lectured to 'larger audiences than he had on previous visits.' Later in the year he made yet another visit to Liverpool, where, in week commencing the 5th November it was reported that he spoke to respectfully attended audiences. The boxes had been full on all three evenings; indeed, on the first two evenings there were many who had wished for box tickets that had to be content with those for the pit. The report made the comment that despite the relative frequency of his visits to Liverpool, he would always be well-received and supported whenever he chose to return, 'for he is truly a practical man and a clever and indefatigable lecturer. He has a voice equal in compass to the largest building, is quite at home even in the details of the science, and often relates an anecdote with a pleasing humour and much force of expression.' The article closed by announcing that 'Mr. Holden is about to visit Warrington and Wigan.'

A report of his three-day visit to Warrington at the very end of November again talked of respectable attendances, and ended by saying that he was proposing to visit 'some of the neighbouring towns,' but without being specific.

The *Northern Star and Leeds General Advertiser* of the 27th October 1838 gave details of Moses being in the midst of a pair of three-day courses of lectures in Rochdale, a place I have not encountered him previously. Ironically, the only theatre that I can trace that was open at the material time was one called 'The Theatre', and that had been converted from a Methodist Meeting House together with twelve cottages.

On the 10th of the following month, the *Leeds Mercury* carried an article about Encke's Comet, that had been observed by Moses Holden on the 8th November. It gave details of its past and future movements and gave guidance as to how to find it in the heavens. The article added that Moses was currently lecturing in Halifax, a town linked by canal with Rochdale, certainly as far as Sowerby Bridge.

We have already had a look at the receipts and outgoings in respect of his 1839 lectures, which were delivered in mid-May, but prior to that, on the 6th April, it was reported that he had been engaged by two societies in Wigan to deliver his courses of lectures, 'which will probably occupy him for about a month.' Unfortunately the identity of one of the two Wigan societies is unknown, but one was the 1825 founded Mechanic's Institute, and was held in the St. George's school-room.

The autumn of 1839 saw an extremely busy period, when Moses delivered a course of lectures on Optics to the Darwen Mechanic's Institute on the 25th, 26th and 27th of September, and a recapitulatory lecture on the 1st October, in the Trinity Church school room, followed three weeks later in neighbouring Blackburn, where he lectured on Astronomy on the 21st, 23rd, and 24th, and then on the 29th, 30th and 31st October, on Optics, both in the theatre of that town.

By Particular Desire.

THEATRE, BLACKBURN.

FOR THREE NIGHTS ONLY.

OPTICAL LECTURES,

ILLUSTRATED BY A GREAT VARIETY OF

TRANSPARENT SCENERY, OPTICAL INSTRUMENTS, &C.

MR. HOLDEN

Returns his sincere thanks to the Ladies and Gentlemen of Blackburn and its Vicinity, for the patronage with which he has been favoured while delivering his ASTRONOMICAL LECTURES, and takes the liberty to recommend to their notice these LECTURES which will be instructive, entertaining, and purely scientific; from which both youth and age may derive advantage and pleasure. The entertaining part, with the Phantasmagoria, is to expose and explain how a variety of strange effects are produced by this instrument.

THE LECTURES TO BE DELIVERED

On Tuesday, 29th, Wednesday, 30th, and Thursday, 31st of October, 1839.

SUBSCRIPTION FOR THREE NIGHTS, BOXES 8s., PIT 5s. 6d.,

SINGLE NIGHT,

BOXES 3s., PIT 2s., GALLERY 1s.

Tickets to be had of Mr. MORRICE, Mr. WOOD, and Mr. WALKDEN, Booksellers; and of the LECTURER, at Mr. Brook's, Livesey Street.

Doors to be opened at Seven, and the Lecture to commence at half-past Seven o'Clock each Evening.

THE THEATRE WILL BE WELL AIRED AND VENTILATED.

SYLLABUS.

LECTURE 1. TUESDAY NIGHT.

Definition of Optics—Light—its prodigious velocity—refrangibility and reflexibility—how decomposed—colour—its property to carry images into a dark chamber through a small aperture—camera obscura—and lucida—subjects painted on the eye in the same way—perfect vision—causes of imperfect do.—Plans of the human Eye, shewing distinct—long—and short sight—Optometer or Eye meter—Spectacles or Lenses of different kinds to help imperfect sight—Eyes of animals, insects, &c.—The Solar Microscope, with plan and specimens.

LECTURE 2. WEDNESDAY NIGHT.

Optical glasses, or Lenses—the practical way of forming these—selection of Glass—edging it—Tools for grinding and polishing—how a number are ground at once, with specimens, &c.—Instruments formed of Lenses—Microscopes—Single, as Leeuwenhoek's—Wilson's—Botanic, &c.—Compound, as Culpepper's—Martin's, &c.—Microscopic Micrometers, &c.—Magic Lanthorn—how to paint for, and form Transparent Medallions to exhibit—with plan and specimens.

LECTURE 3 THURSDAY NIGHT.

Telescopes, Galileo's—Opera Glasses—Sir Isaac Newton's reflecting Telescope—Gregory's—Cassegrain's—Dolland's Achromatic—Day and Night Glass—Herschel's Telescope—Stanhope's, with plans of the whole—Telescopic Micrometers, &c.—Phantasmagoria—Various specimens of painting for—and the strange effect produced by these.

W. H. MORRICE, PRINTER, MARKET-PLACE, BLACKBURN.

Handbill

Immediately prior to a letter being received from the Rev. H.W. McGrath, Moses had attended a meeting of the Preston Branch of the British and Foreign Bible Society in the Theatre Royal, Preston. It was the twenty-fifth annual meeting of that group, and the Theatre was filled, and in some parts was overflowing. Moses' friend and neighbour, the Rev. T. Clarke took the chair for the meeting, which was addressed by a number of reverend gentlemen and Moses Holden, plus a representative from the Parent Society. During the proceedings a collection was made, and contributions totalling over twenty pounds were received, more than had ever been collected before at such a meeting for several years. Although Moses was a regular attendee at meetings such as this one, very few of them were recorded.

The letter from the Rev. McGrath to Moses was, it would appear, a means of introducing himself to him, saying that he had heard that it was Moses' intention to visit Manchester again to deliver a course of lectures on astronomy. He had been friendly with the Rev. Roger Carus Wilson, Vicar of Preston, who by that time had died, and had learned a lot about Moses from him. In the event he delivered two courses in the Theatre Royal, with the second (and positively the last, according to the advertisement) course being given on the 30th, 31st March, and the 1st April. The advertisement in the *Manchester Courier* accompanying the lectures stated that his transparent orrery now measured thirty-five feet in diameter. The first course was delivered during week commencing 23rd March.

Before we move on to the next decade, I'd like to mention an incident that found its way into the *Preston Chronicle* in 1893. It was headed 'Chess in Preston', and related to the year 1840. At around that date there had been a movement among some of the members of the Institute for the Diffusion of Knowledge, to introduce the game of chess as an additional attraction; but some of the elders on the committee were strongly opposed to the notion. Among the objectors, the article recorded, were the late Moses Holden, astronomer, and John Hamer, manufacturer, who were equally opposed to the introduction of novels, and even of "Punch," declaring that such publications did not comprise any 'useful information'.

However, the matters were debated, with the outcome leaving 'Old Moses' and his colleagues without a leg to stand on!

Chapter Eight

1841 – 1850

Following the comment at the end of the last chapter, I think that there can be no better way to start this one other than by looking at the comments made by Professor John Tyndall (1820 – 1893). Tyndall was an Irish-born, brilliant experimental physicist and science educator. In a book entitled 'New Fragments', Tyndall wrote, "In 1842, and thereabouts, it was my privilege to be a member of the Preston Mechanic's Institute (sic) – to attend its lectures and make use of its library. A learned and accomplished clergyman, named, if I am right, John Clay, chaplain of the House of Correction, lectured from time to time on mechanics. A fine, earnest old man, named, I think, Moses Holden, lectured on astronomy, while other lecturers took up the subjects of general physics, chemistry, botany, and physiology. My recollection of it is dim, but the instruction then received entered, I doubt not, into the texture of my mind, and influenced me in after life."

These words were spoken by Tyndall as part of a speech delivered by him when he opened the sixty-second session of the Birkbeck Institute, London, in 1884, and they were repeated by the *Preston Chronicle* in the same year when he was called on to deputise for the otherwise engaged Earl of Lathom who had been due to distribute the prizes and certificates at the Harris Institute the same year. We will learn shortly that the Harris

Institute opposite Avenham Colonnade in Preston, had originally been constructed specifically for the Preston Institute for the Diffusion of Knowledge

In 1842 Tyndall came to Preston as a twenty-two year old member of the Ordnance Survey Party, a position he saw as a stepping stone onto his intended career as a civil engineer. During the course of the speech to the Birkbeck Institute, he referred to 'taking a line of levels between Howarth and Keighley' on behalf of the railway companies, and it was probably in a similar capacity that he was engaged in Preston. By 1843 he had moved on to new pastures.

Returning to old pastures, however, Moses made another visit to the theatre in Lancaster in 1841. It will be interesting to see whether there's any diminution of his lecturing activities now that he's reached his sixty-fifth year. In the August of that year the *Lancaster Gazette* carried a notice that was far more personal than any I had seen previously, and perhaps gives an insight into the way in which he was viewed by his audiences. It read, 'We find that our old friend and neighbour, Mr. Holden, the astronomer, is about to pay his periodical visit, and we doubt not, that if for old acquaintance sake only, our townspeople will renew their patronage with accustomed kindness.' Regard of that nature has to be earned.

A further example can be seen the following year, his 1842 Triennial Lectures in the Preston Theatre Royal at the end of April, drawing press comments such as, 'On no previous occasion was the attendance more numerous or respectable than on the present.' On the first evening, when he presented himself to the audience, the warmth with which they greeted him must have convinced him that the esteem in which his fellow townspeople held him, had in no way abated. The *Preston Chronicle* recorded that in their opinion that if anything it had increased.

Things must never stand still, however, and the paper noted that much of the scenery used in these lectures was new, 'and, we believe, in all cases the result of the lecturer's own labour and skill. The first scene gave a clear demonstration of the revolutions of the Sun and earth producing night and day, and the changes of the seasons. He included ten telescopic views of the Moon, 'of the extra-ordinary size of nine feet diameter,' which were very beautiful and effective.

During the course of these lectures, Moses alluded to the effects of time as he experienced them in his own person, and expressed a fear that it might be the last time that he would have the pleasure of appearing in this manner before his friends. In the event it was a further ten years before his final appearance here.

Many years ago I became aware that in 1842 an oil painting of Moses Holden was produced by Preston artist, Charles Hardwick. Hardwick was a contemporary of William Archimedes, and the son of a licensee in Preston, William Hardwick at the Grey Horse and Seven Stars in Fishergate. Charles's brother, also William, later became the landlord of the same inn. Whilst undoubtedly a talented artist, he went on to be better known as an author of many books relating to Preston's history.

An advertisement in the *Preston Chronicle* in May 1842, placed by Hardwick, made no reference to an oil painting, but 'A portrait of Moses Holden'. He went on to announce in the advert that 'in compliance with the wishes of several respectable parties, it is his intention to publish a portrait of that gentleman. It will be executed in the best style of lithography, by a first-rate London Artist, and will be the same size, and as nearly as possible, a facsimile of the original picture, drawn by Mr. Hardwick, expressly for this purpose'. Prints and proofs were offered respectively at five shillings (25p) and seven shillings and sixpence (37½p), to subscribers who could inspect the original drawing at the *Chronicle* offices.

Whether this refers to the oil painting which was said to have been completed, or something totally different, I do not know, but on August 20th 2014, perhaps a dozen years after my first connection with Moses Holden, I received an email from a stranger who introduced himself as the three-times great grandson of Moses Holden. He was the first living member of the Holden family I had encountered. His home is in Hertfordshire, and he had chanced upon my interest in his ancestor. In a subsequent communication he offered me the information that on his living room wall was an original oil painting of Moses Holden, executed by Charles Hardwick.

With the velocity of light – that's technically misleading, but I'm sure you will understand my excitement – I presented myself at the door of Moses' descendant, and was promptly introduced to a much younger Moses Holden than I had seen images of previously. What a day! What a moment! One hundred and seventy years separated me from the day it was painted, a time when it was described by those who knew him as 'a perfect likeness'. Yet there was more to come. On the adjacent wall was a large pencil drawing of a twenties-something William Archimedes Holden, again drawn and signed by Charles Hardwick, and clearly depicting the 'gentleman of leisure' that was Archimedes' preferred description of himself. Next to the drawing of William Archimedes was a water-colour painting of Mary Blinkhorn, the girl that William married in 1845. In the Holden family, from the date of the painting and probably earlier, she has been referred to as 'The Belle of Bolton'. She was the daughter of a mill owner in Bolton, and when his mill was demolished later in the 19th century, the site became the long-time home of Bolton Wanderers Football Club, Burnden Park.

Moses Holden 'Taken from 1842 oil painting'.

Painted in 1842, this image by Charles Hardwick
shows a 65 year old Moses Holden. Whether some
artistic licence has been taken by Hardwick to show a
dark-haired image isn't known, but the painting was
said to have been a good likeness of Moses.

I must add, whilst talking about images, about four years later in 1846, some portraits were taken by Mr. Eastham of Preston, which were said 'to be the equal of the best specimens we have seen compared to the work carried out by Mr. Beard of London'. One of the portraits was a miniature of Moses Holden, 'which gives us that able philosopher to the life. It is *the* portrait of our eminent townsman, and ought to be handed forthwith to the engraver, to be multiplied by hundreds for the gratification of his many admirers'. It is believed that none of these images remain in existence.

In September 1842, Moses, a Life Member, attended an annual meeting of the British Association for the Advancement of Science in the Theatre of the Royal Institution, Manchester, where he read and explained a paper concerning 'a simple method to calculate the decimal part of the sine or tangent, below a second of a degree, to the 10,000th, or 1,000,000th part of it, to be used to find the distance of the stars'.

Among the notable people present to hear the paper was the Very Rev. Dr. George Peacock, Dean of Ely, a Scottish mathematician, who took the chair. Also present were internationally recognized individuals, for example Professor Friedrich Wilhelm Bessel, of Konigsberg, Germany, who Moses cited in the paper he delivered, Sir John Herschel, the Old Etonian, English born son of Hanover native Sir William Herschel, who, apart from being an astronomer, was also a multi-faceted scientist and experimental photographer, Sir William Hamilton, who held the honorary title of Royal Astronomer of Ireland in Dublin; Professor Moritz von Jacobi, the inventor in 1838, of electrotyping, and many other distinguished persons who were probably more qualified, and yet intellectually, Moses was the equal of most.

Ticket for the Association for the Advancement of
Science when they met in Manchester in 1842.

Remaining briefly with the 1842 Annual Meeting of the British Association, one of those who presented a paper at that meeting was William Henry Fox Talbot, the man said to be the inventor of modern pre-digital photography. His paper was devoted to the improvements that had been made in connection with telescopes, but an observer from the Fraser's Magazine commented in the September 1842 edition, that *'he gave it (his presentation) in a see-saw, hesitating manner; making one quite fidgety to hear him. He was proposing that the speculum be made by the electrotype process, and was trying to explain a mode of obviating the difficulties arising from weight, when reflectors are so large (two or three tons for instance).'* He continued, *'I could scarcely understand his meaning, though he hammered from first to last, a good part of an hour.'*

Our observer continued, *'At the conclusion, up stood a plain-looking man, a self-taught itinerant lecturer on astronomy.'* He is here describing Moses Holden, and he continued by saying that Moses made his way to the large blackboard situated behind the chairman, saying, *'with your leave, Sir, I will just chalk it – I'll not detain you a minute.'* The chairman's appearance was said to be far from inviting.

Undaunted, Moses completed his sketch, and within a minute had made everything plain to his audience. The audience, in turn, immediately showed their approval in an unequivocal manner, with Moses adding, as he was bundling his way back among the benches, *'Now, I've not been long.'* The commentator added that this comment attracted an outburst of laughter, evidently at the expense of the long-winded attempt of his predecessor. *'I could see it was not relished among the savants.'* Our friend from Fraser's Magazine went on to describe Moses as having a *'quaint, brusque, and at times authoritative manner which often caused amusement to his audience.'*

Having read the title of the paper that Moses delivered in the previous story, one would need little convincing that the astronomer deals in exactitudes. That fact again became apparent in July 1842 when he reported the occurrence on the previous day of a solar eclipse. It had occurred in the early hours of the morning between 4.43.30'a.m and 6.34.10'a.m, but the start of it had not been seen in Preston because the sun was at that time covered in a cloud. Moses called in at the newspaper offices after the type had been set, to tell them that the watch he had used for the timings was seven and five-tenths of a second fast, so the times given need to be adjusted accordingly! An addendum was added.

'Once the cloud had cleared', Moses wrote, *'nearly one quarter of it was covered, and a beautiful twilight skirted the edge of the dark moon for some time. There was a great vibration all along the dark edge of the moon that was projected on the sun, like the view of an object seen through the heat of a brick kiln; this continued to the end. At the mid-point of the eclipse, about nine parts of the sun out of twelve were covered. All the time to the end, the mountains of the moon were seen projecting some elevated cones, and others long, and sloping off each way, which made the valleys appear.'*

We have already seen that Moses made lenses for George Horrocks and others, but in this article he wrote regarding his observations of the eclipse, 'The instrument used was a fine achromatic of Dolland's, more than five feet focal length, with a treble object glass nearly four inches in diameter, belonging to G. Horrocks, Esq,' but where the observations were made isn't clear.

Moses' calculations for this type of event were not solely for Preston, or even this country, and it has to be remembered that he did receive information from the Greenwich Observatory, but he felt it worthwhile to add the following information to his article for the benefit of his Preston readers:-

'This eclipse has been a notable one in the south of Spain and France, being total, and the darkness lasting three or four

minutes. It has passed over Marseilles, Genoa, Turin, Verona, Vienna, and, passing on in a line curving south, in the form of a bow, through Asia and on to China, and, passing Nankin, entered the Pacific Ocean, and passed along it for more than two thousand miles. The line is 100 miles broad and 9,500 miles long, where total darkness has passed over.'

Moses' desire to engage people in a more rounded approach to his subject, rather than being content with a limited area of it, manifested itself again later in that year, when he delivered a series of lectures on the History of Astronomy, in the lecture room of the Institution. The lectures consisted of an elaborate review of the early history of astronomy, as far back as the time when man first appeared. He ventured the suggestion that astronomy must have been the first branch of science with which man engaged, but I'm not sure where the evidence for that is. In fact, he said that whatever the ancient philosophers discovered was worth little or nothing, so there probably is no evidence.

He then reviewed the amount of knowledge that was acquired by the Chaldeans, the Egyptians, and the Indians. He used *'some excellent diagrams that were drawn by my friend, Charles Hardwick, the artist.'*

In the second lecture he examined the knowledge acquired by the Chinese and the Greeks, but that of the former amounted to very little. The Greeks, particularly through Thales and Pythagoras, had discovered many facts and principles in the system of the universe, and were, in fact, acquainted with some facts that had been generally supposed to be only of modern discovery. Both the lectures were *'crowded by highly respectable and attentive audiences'*.

The beginning of 1843 saw the novelty of a letter that Moses wrote to the *Preston Pilot* being repeated verbatim in both the *London Standard* on Monday 16th January, and the *Bath Chronicle and Weekly Gazette* of Thursday 19th January. The original, that appeared on the 14th January in Preston related to the severe weather being experienced, with Moses recording that *'the barometer was lower than I have ever seen it any time since I began to observe. At 11.30a.m on Friday 13th, it stood at 27.93. This is very uncommon, and particularly so at the low elevation which my house is above the level of the sea, which cannot exceed 80 feet. As to the wheel barometers, some of them have actually gone round the dial to 'very dry.' I expect we shall hear of a great storm that has happened somewhere, either this morning or last night.'* It was the widely-ranged weather interest that created the attention of newspapers so distant.

This letter to the *Preston Pilot* is an example of how Moses earned the erroneous reputation of being a foreteller of the weather, with country-folk in particular making comments like, *'It's going to rain for the next seven days, 'cos Moses Howden 'ad sed soa'.*

A report in the Preston Chronicle in April 1843, spoke of a Comet that I would describe as an astronomer's Comet. It was one that the layman would have difficulty in identifying and tracking, for it was heading directly towards the earth with the tail behind it. Moses had details from Sir John Hershel that enabled him to locate it, and he later received confirmation from William Rogerson at the Greenwich Observatory. Rogerson added to the communication though, having had some correspondence from the West Indies, where the views had been totally different. The letter read:

'A grand phenomenon - a most splendid Comet - started into view here four or five nights ago. The head, or nucleus, becomes visible about half-an-hour after sunset; and, though the moon is nearly half, yet it appears bright and well-defined. The tail is about one degree and a half broad, in its widest part, and extends to the length of about thirty degrees.'

Antigua, West Indies, March 8th 1843

Moses' appetite for lecturing appears to have waned little, despite his now 67th year. In fact, his stamina was tested to the extreme when he presented nine lectures in eleven days in the Liver Theatre, Liverpool. They consisted of back-to-back courses of his familiar three lecture astronomical performances, from Wednesday 20th March until Saturday 30th March. This is the only time that I have found him giving a trio of courses, and the only time when he has lectured on a Saturday evening. It is thought that he gave an immediate repeat of his course in Manchester on one occasion, and is no doubt indicative of the far larger populations of Liverpool and Manchester, and particularly the target audiences of upper middle class and professional men and women, compared with the other places he visited.

THEATRE ROYAL, CHURCH STREET,

LIVER THEATRE.

OURANOLOGIA;

Or, THE HEAVENS DISPLAYED, in

A COURSE OF ASTRONOMICAL LECTURES

(To be repeated the following week), Illustrated with a most

BEAUTIFUL CEASTRODIAPHANIC,

OR GRAND

TRANSPARENT ORRERY,

TWENTY-FOUR FEET IN DIAMETER, with

SUPERB SCENERY; THE SUN, MOON, PLANETS, AND STARS.

Shining as they do in Nature, enlightening all the place.

Mr. M. HOLDEN,

PRACTICAL ASTRONOMER,

Grateful for past favours, respectfully informs the Ladies and Gentlemen of Liverpool and its Vicinity, that he will deliver three COURSES, (of THREE LECTURES each), familiarizing the Sublime Science of ASTRONOMY, by first describing and then exhibiting the same on this Splendid Machine, illustrating the grand Operations of Nature, as displayed in the Motions of the Heavenly Bodies, with all the Phenomena arising therefrom, such as Day and Night—the Rising and Setting of the Celestial Orbs—the Vicissitudes of the Seasons—Full and Change of the Moon—Eclipses—Transits—Comets—Constellations—Nebulæ, &c., with an immense quantity of Splendid Scenery, consisting of Telescopic Views, &c., surpassing anything of the kind ever exhibited in this kingdom for Correctness; drawn by the Lecturer from Views by Telescopes of great power, made by himself, which Drawings he will have the pleasure to exhibit in this Course to a Scientific and Discerning Public, whose Patronage *alone* he would humbly claim.

THE FIRST COURSE OF LECTURES TO BE DELIVERED

On Wednesday, the 20th, Thursday, the 21st, and Friday, the 22nd,

THE SECOND COURSE

On Monday, the 25th, Tuesday, the 26th, and Wednesday, the 27th,

AND THE THIRD COURSE

On Thursday, the 28th, Friday, the 29th, and Saturday, the 30th of March, 1844.

SUBSCRIPTION FOR THREE NIGHTS OR COURSE.

Boxes .... 8s. 6d. | Pit .... 5s. 6d.

SINGLE NIGHT.

Boxes .... 3s. 0d. | Pit .... 2s. 0d. | Gallery ...... 1s. 0d.
Schools & Children admitted at Half-price. Also Family Tickets.—Apply to the Lecturer.

TICKETS may be had of Mr. CANNELL, Stationer, Castle Street; Messrs. WILMER & SMITH, Booksellers, Church Street; Mr. ABRAHAM, Optician, Lord Street; at the MERCURY and COURIER Offices; of Mr. M. LEICESTER, Tea Dealer, St. Anne's Street; and of the Lecturer, No. 22, Pleasant Street, Clarence Street, Liverpool.

Doors to be Opened at Half-past Six, and the Lecture to commence at Seven each Evening.

A COURSE CONTAINS THREE TICKETS, ONE FOR EACH EVENING.

Mr. H. has published a Third Edition of his beautiful concise ATLAS of the HEAVENS, in 20 Maps 12mo., whereby all the Stars visible in the Latitudes of Britain, may easily be known, with their names, and may be had at his Residence, and at the Lecture. Price 7s. 6d.

[TURN OVER

Liver Theatre Handbill.

SYLLABUS.

J. F. CANNELL, PRINTER, 18, CASTLE STREET, LIVERPOOL.

Rear of the handbill,

After the first two evenings of his mammoth Liverpool endeavour, an article appeared in the *Liverpool Mercury*, reporting that they were glad to see such a numerous attendance, amongst which were many from the top schools in Liverpool and its environs. Once again, the newspaper extols the worth of *'one of the most splendid orreries ever exhibited, and of hearing one who is well-known, not as a speaking lecturer only, but as a practical astronomer, especially as this is most probably the last time Mr. H. will be able to visit Liverpool.'*

The two comments, first in Lancaster and then in Liverpool, about the possibility of it possibly being Moses' final visit is a plausible story that is being given to the press, and probably intended to boost attendance figures.

However, that was probably a part of the marketing strategy, but Moses still had a long way to go, and within three weeks he appeared once again in the theatre in Warrington. The article reviewing the course was again generous in its praise, and particularly that for *'the large illustrated scenes, drawn by the lecturer himself, consisting of telescopic views of the sun, moon and the earth, and a splendid transparent orrery.'* It also noted that the Warrington Mechanic's Society had come to some arrangement whereby their members could attended the course free of charge.

Moses connection to George Horrocks and his insistence on preciseness were both evidenced at the time of an eclipse of the Moon on the 31st May 1844. Moses had had a letter from the Greenwich Observatory informing him that visibility had been poor in London, and only a fraction of the eclipse had been seen just as it was ending.

On the contrary, Moses reported that for us, in Preston, *'the moon rose in splendour, with her full red orb, the air being beautifully clear; and it was soon evidenced that the penumbra (partly shaded area) occupied the south-east side of the Moon.'* He continued by saying that, *'The time I got correctly by repeated* observations, and I was pleased to find that Mr. G. Horrocks, Esq., had the true time likewise, so that we could depend on it to the greatest nicety.'

Moses went on to say, 'I was at the telescope, and Mr. Horrocks had a chronometer in his hand. I spoke the instant the eclipse begun. He said the time was 8 hours 58 minutes 14 seconds. The time calculated for Preston was 8 hours 58 minutes 13 seconds. A second of time must have elapsed before I saw it; therefore, it took place the instant predicted.' In other words, the calculation was correct, but he hadn't announced it to Mr. Horrocks quickly enough. One second of time was obviously vital to Moses.

The rest of 1844 continued with plenty of evidence that Moses was busy with his astronomical observations, but there would seem to have been no further lecture arrangements. There are also odd records of him giving sermons, such as the two that he delivered at the Sabbath School, Abbey Mill, Withnell. In the afternoon he spoke from Ephesians, Chapter 3, verses 14 – 19, a passage that constituted a prayer for the Ephesians. In the evening, he spoke from Exodus, Chapter 34, verses 6 and part of 7, proclaiming the compassion of God, his slowness to anger, his love and faithfulness, and his forgiveness of wickedness, rebellion and sin.

Later in the year, Sunday 13th November, he gave two sermons at the Wesleyan Methodist Association Chapel in Clitheroe, in aid of the Sunday school, and the two collections amounted to a remarkable twenty nine pounds fourteen shillings two and a half pence (£29.71p).

In the meantime, on Monday 2nd September, he displayed his continuing interest and support of the Temperance Movement when he attended, and spoke at, a Temperance Tea Party in the Saul Street Primitive Methodist Chapel; a building that still exists in the form of Ashlar House, the Masonic Hall. A stone memorial can be seen referencing the Primitive

Methodists high up on the south-west facing wall of the building, and can be seen from Ringway.

In late July Moses wrote a letter to the *Preston Chronicle* announcing two events; firstly the inferior conjunction of Venus with the Sun, and secondly a very faint Comet that he had seen on two successive nights, for which he gave details, but it was a week later, due to poor visibility that he was able to report on its progress and position. In total he wrote to the paper for four successive weeks, enabling a number of people to witness it before disappearing.

A month later, however, he again wrote to the paper about another Comet. On the 20th September it had only been seen from positions on the continent of Europe, but he had calculated and gave the position it would be seen in on the 21st in this country. He also reported that the planet Jupiter was closer to the Earth than it would be during the next twelve years, and consequently looked much larger than usual. Moses commented that with a power of 90 times diameter, the planet now looks much larger than the full moon (with the naked eye). Three weeks later, on the 12th October, he again wrote to the *Chronicle*, but reported that his health and the weather had prevented him from Comet-hunting during much of that period. Both had now improved and he gave its current position, with a follow-up on the 19th October when, in addition to its position Moses also recorded that the Comet was getting further and further from the Earth, and will disappear.

In the Manchester Courier and Lancaster Gazettes of the 28th September 1844, an auction sale was to be held in the salerooms of Thomas Wren, auctioneer and upholsterer. Part of the sale was dedicated to the residue of the collection of pictures, engravings, water-colour drawings, and articles of furniture of Sir Peter Hesketh Fleetwood. One of the lots was *'a large and powerful microscope, with a number of powers, two of* which are French achromatic, and one of Ross's one inch achromatics, by Moses Holden.'

The year's second Eclipse of the Moon occurred before the end of November, with Moses computing figures for both the Liverpool and Preston newspapers. I referred to this event in the last chapter, when Moses commented on the unreliability of the public clocks in Preston, but he also made a comment of general interest, which he will have determined from his calculations. He informed readers that *'When total darkness begins, the moon will be vertical, just over the east entrance of the Great Sandy Desert of Sahara, Africa, in 21° of north latitude; and just as that darkness ends, the vertical moon will leave the tremendous desert of Sahara, that it has traversed across while immersed in the earth's shadow and enveloped in this awful gloom.'* Information such as this requires a consideration of the probable scant geographic knowledge of the majority of the newspaper's readers. It must have added to the overall mystery of the event?

Conditions for the viewing of this eclipse were very similar to the one in May, with observers in London seeing only the final stages of it. Again, in Preston, visibility was good. He then went on to make a statement that led to some entertaining criticism in the Preston Chronicle at each side of the New Year. He wrote:

'Some persons have asked why the moon was so visible and red all the time of the total eclipse. My answer is, that if the earth had no atmosphere, the moon would disappear when totally eclipsed; but its appearing when it is so, is owing to the rays of the sun passing through the earth's atmosphere, and being bent inwards by refraction to the very centre of the earth's shadow; and as the red rays are refracted with the greatest ease, they are conveyed to the moon, and reflected back to the earth. During the total darkness of the last two eclipses of the moon, I could see the principal marks upon it through a telescope. This is nothing new. I do not remember one total eclipse of the moon, when that body has entirely disappeared; though it is said this

happened in the total eclipses of the moon in 1601, 1620, and in 1642.'

I suppose that it has always been the case that those who are regarded to be the leaders in their field, will, at some point, become the target of criticism. Constructive examples of such comment will generally be directed privately through the correct channels, but those who *think* they know more seem to enjoy sharing their grievances in the public eye, and in the 1840s that meant the Letters to the Editor section of the local newspaper. Such was the case around the turn of the year from 1844 to 1845.

A reader who called himself E.D. had, after reading an account of the eclipse by Moses Holden in the *Preston Chronicle*, written to that paper's opposition, the *Preston Guardian*, in which he detailed his disagreements. In a one thousand word response to that criticism, Moses wrote again to the *Chronicle*, in which he said, "Now, Sir, let this youngster (for I cannot imagine him being a man) peep from his gloomy shadow and look at this, and see if I have not swept away his *cannot*, for the doctrine is more than borne out." Moses went on to quote from Sir David Brewster, an expert on the science of optics, before finishing with another strike at his letter-writing foe, by adding, "Now, Sir, I shall not by enchantment, attempt to deliver this youth from his gloomy shade, although I think this letter will have cast some light on his understanding; I would advise him, before he writes again on subjects of which he is perfectly ignorant, that he should go to Jericho, and tarry there till his beard grows!" Astonishingly, E.D. wrote again with further criticism, but Moses ignored it.

Although we will see evidence that Moses gave a series of lectures in Preston in 1845, in the main he limited himself publicly to corresponding with the newspapers, with the first example being an account of a Comet at the end of January. He described it as *'a small and faint object at present'*, but after predicting its position on the 1st February, never returned to the matter.

Moses' next letter to the *Preston Chronicle* on the 3rd May, related two events, and is worth repeating in its entirety:

ECLIPSE OF THE SUN AND TRANSIT OF MERCURY

Sir, In the ensuing week –commencing 5th May, should the weather prove clear, those who have pleasure in viewing celestial phenomena may have their taste gratified by observing not only a solar eclipse, but a transit of the planet Mercury.

On Tuesday morning, the 6th May, there will be a partial eclipse of the Sun. The eclipse begins at –

Preston, true mean time 8hr 20min 12 secs

Greatest obscuration 9 31 0 The eclipse ends 10 37 57

Dividing the sun into 12 equal parts (or digits), five of these and nearly a half will be covered; the moon will make the first impression on the solar disc 41 degrees, to the right of the sun's vertex or upper point. This eclipse will be larger with us than at Greenwich, and it will be central and annular at the north polar seas.

TRANSIT OF THE PLANET MERCURY OVER THE SUN

The planet Mercury, that was seen so bright in the west for a few evenings about the 17th of April last, will be like a round black ball on the sun, on Thursday afternoon, the 8th of May. The first contact when the transit begins

Will be 4hr.8mins 18secs Middle 7 24 14

Sun sets at Preston 7 44 10 Transit ends 10 40 1

Mercury makes its first impression on the upper part of the sun to the left hand. We shall see the transit (if the air be clear) much longer than they will at Greenwich.

13, Jordan Street, May 2nd 1845 Yours. &c. M. Holden

Such are the vagaries and unpredictability's of the weather in ones astronomical observations, that on the following Saturday, Moses reported in the *Chronicle* that his timings relating to the eclipse were at fault by two seconds – a human error – and that the afternoon of the 8th had been cloudy, and at no time was the face of the sun seen.

Early in September, Moses announced that he would deliver his Triennial lectures at the Theatre Royal once again. They took place there on the 15th, 16th and 18th of that month, to audiences that were described as very numerous. On this occasion he made a concession to the 'operative members' of the Institute for the Diffusion of Knowledge, of half-price tickets to the gallery, meaning a reduction of the normal one shilling (5p), with the Pit tickets two shillings, and Box tickets three shillings, at the usual price.

The newspaper reported that, *'they were much pained at the wanton mischievous conduct of some in the boxes, who repeatedly threw peas and marbles on the stage when the theatre was darkened; and especially so as this improper conduct was traced to parties who from their standing ought to know better. We trust, for the credit of our town that such conduct will not be repeated'*. The *Preston Guardian* of the same date, 20th September, elaborated slightly on that by reporting that *'a number of thoughtless school lads, in the boxes, disgraced themselves, their friends, teachers, and the town, by throwing marbles, peas and other missiles, at the audience, on each evening'*. I think I would have included the parents in the list as well!

A notable exception to the lack of lectures outside Preston, were two sets of lectures separated by just one week, and organised by the Blackburn Mechanics' Institute at the Theatre in Ainsworth Street in the town. The first were three astronomical lectures on the 11th, 12th and 13th November 1845, followed by three lectures on optics just one week later on the 18th, 19th and 20th.

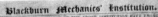

Blackburn Mechanics' Institution.

THE BOARD OF MANAGEMENT OF THE ABOVE INSTITUTION HAVE ENGAGED

MR. MOSES HOLDEN,

Practical Astronomer, of Preston,

To deliver his Course of Lectures on ASTRONOMY; and also his Course on OPTICS, at the

THEATRE, AINSWORTH STREET,

The course on Astronomy on TUESDAY, WEDNESDAY, and THURSDAY, the 11th, 12th, and 13th days of November; and the Course on Optics on TUESDAY, WEDNESDAY, and THURSDAY, Nov. 18th, 19th, and 20th, 1845.

OURANOLOGIA,

Or the Heavens Displayed, on a Course of Astronomical Lectures, Illustrated with a most Beautiful Transparent Orrery or GRAND TRANSPARENT ORRERY, from 19 to 24 Feet in Diameter, to the place may admit, with regard to scenery, the SUN, MOON, PLANETS, AND STARS, Shewing as they lie in Nature, enlightening all the plan.

Mr. M. Holden's

COURSE of THREE LECTURES

SYLLABUS.

LECTURE I. *Tuesday Night.*

LECTURE II. *Wednesday Night.*

LECTURE III. *Thursday Night.*

THE COURSE OF THREE LECTURES ON OPTICS.

SYLLABUS.

LECTURE I. *Tuesday Night.*

LECTURE II. *Wednesday Night.*

LECTURE III. *Thursday Night.*

PRICES OF ADMISSION:

WOOD, PRINTER, BLACKBURN.

The final event of 1845 was the marriage of Moses' son, William Archimedes, to Mary Blinkhorn, the fifth daughter of the late John Blinkhorn, who we met earlier. The ceremony took place at the Parish Church in Bolton, and was held on the 22nd December. They were to produce six children, although two were to die at an extremely early age. They were both girls, and were the fourth and fifth children of the marriage.

William Archimedes' early life isn't recorded at all. If my assumptions about his father travelling around the country between 1815 and 1828 are accurate, then maybe William was home-tutored on the canal barge, for Moses offered his services as a teacher of a number of subjects at various times in his life. There are specific references to him helping his father with his lecturing engagements, but none to suggest that he had any interest in astronomy. In the accounts of his Farewell Lectures of 1852 there is a record of William and Annie, Moses' daughter, being paid for their assistance.

There were a number of suggestions during William Archimedes' life that he had a drink problem, and if ever there is a cause for such thing, it wasn't found in his home. If having a friend such as Charles Hardwick, whose father, as we have seen was a licensee in the town, had any bearing we're never likely to know, but it's not an excuse in any event. Perhaps the most vivid recollection came from photographer Robert Pateson, who, in an article towards the end of his own life, recalled attending one of Moses' lectures. He spoke of Moses calling out to his assistant, William Archimedes, to *'wind Jupiter up and give her a turn.'* In a further, similar story that indicated William had an uncanny knowledge of the closest tavern to the theatre, was when Moses amused his audience by calling, *'Archimedes! Archimedes! Just give the moon a further turn!'* But William *'was in a place where they cared what not of moons!'* In other words he was in the Theatre Hotel bar next door to the Theatre Royal.

Towards the end of the 1840s, William was self-employed as a tobacconist with a shop on Fishergate, but by the November of 1848 he was declared bankrupt. He later found employment as a railway clerk, but failed to settle down. In 1851 he was living with his family in Taylor Street at the foot of Fishergate Hill, and later lived in Garden Street near the railway station.

Again, the sequence of events is far from clear, but in early 1857, a few months before his sixth child, Charles, was born, he went to Australia. Whether it was a Moses-sponsored effort to encourage him to forge a new life, apparently to manage a coffee plantation near Brisbane, presumably in the hope that his family would follow, I have no idea, but the only records of him in Australia have come from Melbourne, some 1,000 miles further south. Eighteen years earlier, in 1839, a contemporary and probable acquaintance of William, John Ainsworth Horrocks, an explorer, had gone to the Melbourne area, where he founded a settlement on the Hutt River, and called it Penwortham. It was in the Clare Valley, and although John's mother was called Clara, it's believed to have been named after the ancestral family home in County Clare in Ireland of the founder of the town of Clare in Australia, Edward Burton Gleeson. An alternative possibility is that it was named after Gleeson's hometown of Clare, in County Armagh.

John Ainsworth Horrocks, born 1818, was the son of Peter Horrocks of Penwortham Lodge, just outside Preston, and later known as Penwortham Hall. John went to South Australia in 1839 on his twenty-first birthday, together with his sixteen year old brother, Eustace. He met an unfortunate end when his elbow was caught by the lurching of his camel as he was loading his gun, causing it to be discharged. He survived for a month, before ultimately dying of gangrene. During that month he ordered that the camel, which had previously attacked humans and a goat, should be 'executed'. It was.

In the absence of evidence, I feel that William was, perhaps, attracted to Melbourne to pay his belated respects to John Ainsworth, particularly if they had been close as teenagers, or maybe the romance of the explorer's story outweighed the prospect of managing a plantation. What I do know is that the ships bound for Brisbane from Liverpool, would have lain adjacent to those for Melbourne or its nearby port of Geelong, and an eleventh hour change of plan implemented. Was he accompanied to the port by Moses? At what point was the one-way ticket purchased? Those are the sort of questions to which, I suppose, we will never know the answers.

The first mention of Archimedes in Australia was in the *Melbourne Argus* of 30th March 1860, in a notice that requested 'William Archimedes Holden, to call at the offices of the *Black Eagle* within 14 days, or your violin will be sold to defray expenses.' The *Black Eagle* was said to be a magnificent ship of the *Mersey* line, and the same request appeared in two or three issues. Why his violin should be held by them isn't clear, for she was only launched from the builder's yard in Quebec during the summer of 1858. Perhaps he'd sailed on a sister ship belonging to the same company.

The next two newspaper mentions were two days apart on the 26th and 28th July, 1862, the first of which gave notice that a man named William Holden had fallen down in Elizabeth Street, Melbourne, close to the Glasgow Arms Hotel. He was taken to the hospital but was dead on arrival there. He had items on him to positively identify him, including two hospital forms that gave the impression that he had been trying to get himself admitted to hospital. He had no money in his possession, and it was said that he had presented the appearance of a man who had lately experienced starvation and cold. It shouldn't be forgotten that this would be Australia's winter.

Two days later, and after a post mortem had been carried out, another report appeared in the *Melbourne Argus*, his full name, including Archimedes was given. It was said at the inquest, by the police officer who had taken him to hospital, that William was well-known to the police as a man who followed no regular employment, but who was a habitual drunkard. The surgeon who examined William's body detailed a list of classic degenerative, alcohol-induced illnesses, and added that a strong smell of alcohol pervaded all parts of his body. The jury returned the inevitable verdict that *'the deceased died from the poisonous effects of alcohol.'*

Moving on from that little aside, 1846 presents us with an interesting mixture of letters to the press and other articles, but no lectures.

A DOUBLE COMET

On Wednesday the 25th February, the evening being fine, I had the pleasure of seeing that wonderful phenomenon, the *Double Comet*.

They are both alike, nebulous, but the south one is larger than the other. They seem to be about six minutes of a degree distant from their centres; and a clear space lies betwixt, so that the largest would go in between. They are moving on side by side. Surely this is the most wonderful thing that has taken place in astronomy, on record.

[Moses then proffered various predictions about times and distances, before adding:]

As this Comet had gone by two names – Biela and Gambart – now that there are two of them, each may claim one for his own.

The evening of Monday 4th May 1846 witnessed the epitome of Moses disparate interests, which also linked the two places that were the most important to him, Preston and Liverpool.

On that evening he chaired a meeting of the Liverpool Seamen's Friend Society at the Corn Exchange in Lune Street, Preston. Its purpose was to promote the objects of that society and the Bethel Union. Moses explained to the not very numerous audience that their object was the imparting of religious instruction to seamen, and providing for them some refuge on their return from sea.

There were two letters from Moses relating to a new planet, the first one on the final day of October, and the second on the first Saturday in November. It was Le Verrier's new planet, which through a common telescope appeared like a star of the 8th or 9th magnitude, which isn't very bright, but that through an instrument that magnified 200 or 300 times it shows a diameter which fixed stars do not have. In both letters he gave the positions it would be found on the four following nights.

A SPLENDID AURORAL ARCH

On Tuesday evening the 17th November 1846, at 6.30pm, I observed a brilliant luminary arch crossing the heavens at right angles with the magnetic meridian; the eastern limb was 25 degrees to the north of the east point, and the western 25 degrees to the south of the west point.

The arch moved southwards, and the centre of it disappeared and reappeared twice before it vanished from our sight.

The north and north-west parts of the heavens were beautifully illuminated by a grand display of the Aurora Borealis, but clouds soon came and swept them all away.

13, Jordan Street,   November 20th 1846

The year 1847 was to see an eclipse of both the moon and later the sun, the sighting of another new planet, another severe barometric depression, and the appointment of a new president at the Institution, but no apparent lectures.

On the final day of March, the partial eclipse of the moon forecast by Moses came to pass, with him reporting in the following week's newspaper that *'I tested my chronometer several times with a fine repeating circle, and had the time true to the second.'* He went on to report that *'Clouds covered the moon at the time of the greatest obscuration, but it was fine again at the end of the eclipse. I was at the telescope, and another person at the chronometer. I called the moment the shadow left the edge of the moon, and this was correct to a second.'*

An interesting paragraph was found in the auction sale notice to be held by Mr. Thomas Shelley Vallet, the auctioneer, on the premises of number 7, Winckley Square, Preston, on Friday and Saturday the 7th and 8th May, 1847. These premises had been built for the Vicar of Preston, the Rev. Roger Carus Wilson, but he had died in 1839, and it's not known whether his successor, the Rev. John Owen Parr, lived there as well.

The items included in the paragraph to which I refer read, 'A superior Magic Lanthorn (sic), with numerous slides, large wooden frame, and 8 − 4 sheet, without seam; an excellent microscope, with additional powers, &c., by Moses Holden, which will be sold about four o'clock in the afternoon of the first day.' It isn't clear whether all these items were made by Moses Holden, but I suspect that they could have been. Don't forget that Moses made a telescope for the Rev. Roger Carus Wilson.

There is probably no relevance, but Mr. Vallet, the auctioneer, had, six years earlier been a cotton manufacturer, living in Bolton Street West, with his thirty three year old wife Alice and six children. That street ran off Pitt Street to a point close to Christ Church where it met Jordan Street and Moses' home. Four years after the sale he was still an auctioneer, but he has a new wife, another Alice, and again thirty three years old, and a new small child. They are living at the Shrewsbury Arms in Oxton, near Woodchurch on the Wirral, and his wife is the innkeeper.

SPOTS ON THE SUN

There are at this time a number of very fine spots on the sun. On Thursday 17th inst, I saw not fewer than fifty, when I viewed the spots through a powerful telescope, four of which were very fine, and were surrounded with large brown margins. The remainder were in clusters, or groups of small spots, crowded together, and appeared very like the perforations of small shot.

13 Jordan Street.   June 18th 1847
Moses Holden

On the 11th May 1848, less than twelve months later, Moses recorded that on that date, using a Dolland's forty four inch telescope, he had seen *not less* than 120 spots on the sun. The largest number he had previously seen was eighty.

THE NEW PLANET "IRIS"

On Tuesday evening last, the 31ˢᵗ August, had the pleasure of viewing the new planet lately discovered by Mr. Hind, and by him named "Iris." I likewise observed it again last night, as it crossed the meridian of Preston at 8hr. 57min. 39secs. It is very bright for so small a planet.

There are now seven planets known, moving between Mars and Jupiter; namely, Vesta, Juno, Pallas, and Ceres, and the other three lately discovered, Astræa, Hebe, and Iris.

The position of Iris for tomorrow evening, the 4ᵗʰ September, will be right ascension, 19hrs 44 mins 21 secs., and south declination, 14° 58′ 50″.

Neptune, the most distant planet of our system, may be seen crossing the meridian of Preston, on the 4ᵗʰ September, at 11hr. 10min. 40½secs. Its right ascension 22hr 3 min 57secs., and south declination, 12° 31′ 25″.

13 Jordan Street. September 3ʳᵈ, 1847.

Moses Holden

ASTRONOMICAL REMARKS.

To the Editor of the Preston Chronicle.

SIR,

As I am now on a visit at the house of my excellent astronomical friend, and your worthy townsman, Mr. Moses Holden, the almost constant wet weather preventing our going out to survey the beauties of nature, which, I am convinced, surround the noted borough of Preston,— I will therefore devote a little time in drawing up a few astronomical remarks, which perhaps may entertain some of your numerous readers.

We live in an age of discovery. Passing by the various comets which have been observed within the last two years, four planets have been discovered, three of which, namely, Astræa, Hebe, and Iris, revolve between Mars and Jupiter, while Neptune, which was first seen in 1846, is almost double the distance of Uranus from the sun, around which it is considered to revolve in 166 years. This planet was in opposition to the sun about the 22nd of August last. On the 17th of September, according to Mr. Holden's computation, its distance from the earth, in English miles, was 2,788,125,000. The right ascension of this planet, on the 25th instant, will be about 22 hours 2 minutes, and its declination 12 degrees 42 minutes, south. It appears, through a telescope, like a star of the eighth magnitude. On Monday last, Mr. Holden and I observed the largest group of spots on the sun's disc we had ever seen. It was then approaching the western limb of the sun; cloudy weather interrupted further observations. This group seemed to extend over one-sixth part of the sun's diameter, and in some places was particularly dark. Another group, not so large as the one just mentioned, was coming in on the northern side of the sun, and will therefore be visible for some days to come.

I am, sir, yours, &c.,

WILLIAM ROGERSON,

(of the Royal Observatory, Greenwich).

13, Jordan-street, Preston,

Sept. 23rd, 1847.

'Rogerson 1847 letter'

It would be reasonable to assume from the contents of the first paragraph of this letter that both men had an interest in natural history, although to what extent is under-recorded. The only other time I encountered such an interest was when he was on the Lincolnshire and Cambridgeshire Fens.

ECLIPSE OF THE SUN

[This account of an annular eclipse is not given verbatim, but includes some of my own comments along the way. He wrote,]

'On Saturday morning, October 9th, we shall have a very large and notable eclipse of the sun. [This is the annular eclipse that I have exhibited and told the time of in all my lectures since 1815].' Annular refers to the bright ring that the obscuration by the moon leaves for the observer. Moses expands on that later in the letter.

[Moses then went on to reveal evidence of his skill in calculating the track of the sun, together with his geographic knowledge, a subject that he also taught privately. He explained that,]

'The first point of land where this eclipse will be seen annular is Cape Clear – the most southern point in Ireland – a little after the rising of the sun. It then passes the southern part of the St. George's Channel, and into Cornwall near Camelford, and across Devonshire, passing over Tor Bay to the English Channel, and on through France and the north of Italy and Greece, and over the Persian Gulf – curving through Hindustan a little to the south of Calcutta, and through to the east side of the Burman Empire, where it ends with the setting sun.'

'All along this track the moon will appear in the centre of the sun, and there will be a bright border surrounding the moon's dark body, and of equal breadth on every side. Like a brilliant ring of light which will be beautiful and glorious to behold.

'At Greenwich the eclipse begins at the rising of the sun. The ring there will be unequal; the north-east side will be much broader than the south-west side, which will be narrow, and like a beautiful bright thread of light.'

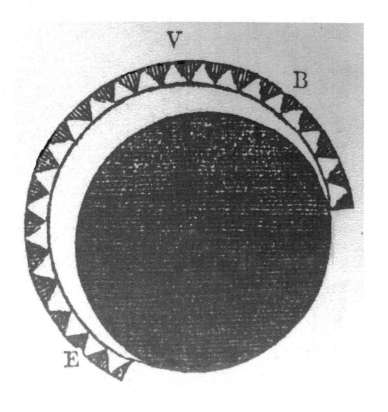

Eclipse of 9.10.1847

V, the Sun's vertex; B, the place where the Eclipse
begins; E, where the Eclipse ends.
13, Jordan Street.   September 29th 1847
Moses Holden

Reports of the eclipse after the event were mixed, with Preston viewers managing to get extremely good views. Unfortunately, Moses chose to walk to Howick House, the home of his friend Thomas Norris who had a well-equipped observatory, but the cloud cover restricted his viewing considerably. His report is worth reading though, in which he wrote:

'The morning of the 9th October proved very fine; I walked to Howick House, thinking that I should have a better view of the eclipse there than in Preston. On my arrival there, a little before sunrise, I fixed the telescope on the place where it would first appear from beyond the hills. As it rose, I saw the first appearance of the sun's edge, and on its getting higher I found the eclipse had made good progress. I then adjusted one of Troughton's sextants on a stand, and measured the dark part of the sun, which at that time, 6hrs 37 mins 38 secs, was 17 mins 30 secs.'

'After this a cloud covered the sun, and I didn't get another observation for an hour and twenty minutes. At that time the eclipse was rapidly decreasing. During the continuance of the eclipse I took the measure of it twenty-one times; and just at the last I took a good three and a half foot Dolland's, and observed the end. T. Norris, junior, noted the time.

'Having computed the time for the end of the eclipse, as near as we could judge of the latitude and longitude of Howick House, the observed end was thirteen seconds later than the computed forecast. The time was taken by a fine transit instrument in the observatory at Howick House.'

The Annual General Meeting of the Institute for the Diffusion of Knowledge was held in early October, a day or two before the eclipse we've just been referring to. It was a time when there was a need to elect a new President of the Institute; a time when they were planning to move from their rooms in Cannon Street. They had, in 1846, laid the foundation stone of a new and grand building opposite the Colonnade in Avenham, the place we knew for generations as the Harris Art Institute. It was to be constructed between 1846 and 1849. It is the fact that Robert Harris adopted this building later in the century, together with the premises in Corporation street that later became the Harris Technical College, that gives us the connecting link to the modern day University of Central Lancashire.

Institute for the Diffusion of Knowledge.

Although the foundation stone was laid in 1846, and the huge stone that is supported by the impressive columns shows a date of 1847, it was closer to 1850 by the time the building was finally opened for use.

At the A.G.M. of the Institute, Moses Holden proposed that John Bairstow, Esq., be elected to the position of president. Bairstow Street, which runs along the eastern side of the new Institute, is a street that takes its name from him; and he was already a well-known benefactor in Preston. Moses told the meeting, "I am happy to inform you that we have consulted Mr. Bairstow on the subject, and he has consented to accept the office."

Moses then added, in a manner which I feel was typical of his forthright approach, "He is a man who is capable of doing them a great deal of kindness, if he had a mind to it. He was a man who had a head; and he had a purse, and plenty in it!" These comments drew a great deal of laughter and applause before he added, "You must not, however, suppose that he has been selected merely for what was in his pocket. His is a man of good, sterling abilities, a gentleman in all his transactions."

It was an interesting speech, for in a similar way to suggesting to his lecture audiences that this may be his last visit, alluding to his advancing years, he again made reference to it here, by saying that he hoped he would see the completion of the new building before he was called away. "You shall," was the reply from many parts of the audience when he mentioned it on this occasion.

Later in the address, whilst talking about the collection of books the Institute now possessed, and particularly those reference tomes that weren't permitted to be taken from the premises, he said, "I still recollect much of the contents of many books which I read in my younger days, and I believe I will never forget them, although I now forget things of a more recent date." Some aspects of growing older never change, it would seem.

Towards the end of the speech, and after mentioning the spaciousness of their proposed new home, and the extra facilities they would be able to offer. He commented that the new Institution would be one of the best lounging places in the town, on a summer's evening. "The observatory placed on the top of it would allow them to have a most delightful and extensive view over the surrounding country. You will be able to see as far as the lighthouse at Fleetwood, with all the windings of the Ribble, and fine landscapes on every side." From that description it would appear not to have been intended as an observatory for astronomical purposes.

I'm sure that when the new building was complete it will have had its own weather station, but for the present, Moses had one at his Jordan Street home, from which he recorded that on the 6th December 1847, his barometer was at its lowest since the Christmas Day of 1821, and even then it was only fractionally lower. He gave several readings of the pressure, and commented that he had little doubt that he would hear of a destructive storm from some quarter or other.

However, he continued, "We have had no storm in Preston. The wind was high, but couldn't be called a gale; and there was no more rain than we frequently experience."

In typical fashion, he then stated that a barometer weighs the atmosphere. He explained that "On Monday night the atmosphere was lighter in weight than its usual mean by a quantity equal to the depth of one inch and four-tenths of mercury surrounding the entire globe."

"The greatest variation," he continued, "of the weight in the atmosphere (as it is with us) is equal to a sea of mercury in depth two inches and a half, covering a smooth globe as large as our earth, on every side." He then made everything clear by adding, "The mean weight of the whole atmosphere is equal to a sea of mercury, covering a large globe the size of our world, twenty-nine and a half inches in depth."

The following year, 1848, followed a similar pattern, with reports of eclipses and spots on the sun being, in the main, the only matters of astronomical interest. There were no lectures again in the year, but October did produce an interesting observation of an Aurora borealis, or Northern Lights. Moses report of the incident in the Preston *Chronicle* of 20th October is worth repeating:

THE AURORA BOREALIS

On Wednesday night, the 18th October, I observed the Aurora borealis; it commenced at 7pm., Greenwich Time. At fifteen minutes past seven the streamers were shooting up in grand style: from half past seven till half past eight there was a broad bright zone crossing at right angles the magnetic meridian from horizon to horizon, and extending in breadth from Delta Draconis North to Alpha and Gamma Pegasi and Alpha Aquilæ South, not less than 60 degrees broad, and of equal light, the streamers dashing from the North and North West nearly to the edge of this zone, and some of them growing brighter and brighter, first in one place, then in another. These were of a brilliant white light.

At forty minutes past nine o'clock the appearance was past all description; the colours were brilliant, particularly the red or lake, to a crimson, and the orange and green set off the scene most gloriously. These coloured lights rose beyond the zenith, and met from the East and from the West overhead.

A little before ten the scene was magnificently grand. About 24 degrees south of the zenith, and near the head of Andromeda, there was a point from which a multitude of streamers emanated and diverged in every direction, eradiating from the centre, and the appearance was very like the spars of a large dome before it is covered. These streamers extended to Saturn, in the south, and to equal distances all round. This strange sight continued some time.

The Aurora lasted till midnight. I have not seen anything like it since 1789.

13, Jordan Street.

Moses Holden

Later in the year, Moses received a letter from William Rogerson at Greenwich, notifying him of an imminent transit of Mercury across the face of the sun. Moses had Rogerson's letter published in the *Preston Chronicle,* and the week after he sent a report of the event which had happened on the 9th November 1848, and which he had witnessed at Howick House. On this occasion it took place in the late morning, and the day was fine and light.

TRANSIT OF MERCURY.

To the Editor of the Preston Chronicle.

Sir,—I have received this communication this morning, from my friend, Mr. Rogerson, of Greenwich; should you give it a place in your paper, you will oblige Yours, &c.,

Preston, Nov. 2, 1848.   M. HOLDEN.

TYPE OF THE TRANSIT OF MERCURY OVER THE SUN, VISIBLE AT PRESTON, IN LANCASHIRE, ON THURSDAY, NOV. 9th, 1848.

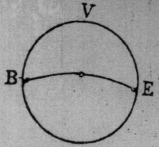

In the above type, V is the vertical point of the sun; B the planet Mercury at the beginning of the transit; M the planet at the middle of the transit; E the planet Mercury at the end of the transit. The above is a natural view, or what a common telescope shows used for land objects. If the type be turned upside down, it will be a cor ect representation through an *inverting* telescope.

The first contact of Mercury with the sun's limb, or, in other words, the transit begins at Preston at 11h. 1m. 42s. in the forenoon; middle, at 1h. 46m. 6s. afternoon. The last contact of limbs, or the end of the transit, takes place at Preston at 4h. 26m. 38s. in the afternoon, true Greenwich time; for that is the time (that is, or *should* be,) used at Preston.

Mercury sets at Preston at nine minutes past four, which is about seven minutes before the end of the transit; so that at the time when Mercury sets, a small space will be perceived between the planet and the sun's edge; the former being entirely on the solar disc when descending below the horizon of Preston.

W. ROGERSON.

Transit of Mercury 1848 letter.

The relevant timings were recorded by Moses and a Mr. Rothwell from London who observed with us all day, and with whose fine micrometer they took measurements. *'Mercury appeared as a beautiful round black ball upon the Sun,'* Moses wrote, *'very different from the spots (on the Sun) that it passed by one after the other; nor did a cloud obscure the sight, until the sun descended, and the trees and hedges hid it from our view.'*

This is the second time that Moses has referred to the appearance of Mercury being that of a 'round black ball'.

In the October of 1847 we learnt of an A.G.M. of the Institute, and the election of a new president. There was a suggestion at the time that financial restraints were causing a slower than hoped for rate of progress with the new Institute building, and over twelve months later things seem not to have improved. In early December, just twelve months later, a Working Man's Tea Party was organised at the Corn Exchange by the committee of the Institution *'to secure the co-operation of all classes in promoting a bazaar'*, that was planned for the spring of 1849. The bazaar, it was hoped, would replenish the coffers of the building fund for that project.

The rationale for the Tea Party was not merely to enlist working men as workers at the bazaar, but also to give a boost to the membership of the Institute itself, and in that respect it was highly successful. The meeting attracted between four and five hundred people, and all but a handful of them were of the operative classes. Food and beverages had been provided for those who attended, and although there were no specifics mentioned it was said that *'the good things could not have been more readily rendered had the recipients been merchant princes, wealthy manufacturers, or rich tradesmen'*; and all had been consumed to the accompaniment of music from the National School Brass Band who had offered their services free of charge for the occasion.

Alderman Birchall, who had been elected president of the Institution for the year, after making an eloquent speech about introducing the finer things of life, and the arts and sciences to the working man, and after the band had played, *Isle of Beauty*, he introduced Moses Holden to the audience.

Moses began by reminding the audience that the last time he had spoken to them there had been little prospect of being in possession of the new building by the summer of 1849, but now that prospect was brighter, and he begged to move the resolution that "This meeting views, with satisfaction, the spirited endeavour of all parties to aid the bazaar for the completion of the new building; and pledges itself, by every means in its power, to promote the bazaar for so laudable a purpose."

In a manner that was unique to Moses Holden, he said, "I feel the satisfaction, the committee feel the satisfaction, and if I am not wrong this meeting feels the satisfaction also." Cheers rang through the Exchange Rooms in approbation. Moses went on to reiterate that the new building would be the finest building there had ever been in Preston, and went on to say how pleased he was to be in the company of so many of the working classes who were clearly willing to give what they could to ensure the completion of the building, so that they themselves could take advantage of the learning facilities that would then be available. "Many of you have it in your power to aid and assist the Institution, and many are doing so – for a number are going from place to place, like begging friars, and not one of them seems to be shame-faced. On the contrary, they are determined to accomplish the object they have in view."

Moses then told the gathering what he called a simple truth, when he was involved with the construction of his celebrated orrery in 1814 and 1815. "When I was engaged in preparing the large orrery that I carried throughout a great part of the kingdom, I had not enough money to complete it in the way I wished." "A gentleman," he continued, "who had been trying experiments with it had promised to lend me £100 or £200 if I was ever short of money. I walked to Bolton to see him and remind him of his promise. I saw the gentleman, but he would not give me a single pound, so I set out to walk back to Preston. As I was walking over the dreary moors, on the New Road, I recalled the fable of the lion and the mouse, and how the mouse, by nibbling at the net which confined the lion, had at length managed to set him at liberty."

He said that he had said to himself at that moment, "I'll nibble, nibble, nibble, till I have accomplished it," and the cheers rang out once more. "In about a year and a half from that time I had accomplished an orrery that had no match in the kingdom, and which I could extend to 100 feet in diameter, and which had been exhibited in the Theatre Royal in Manchester at forty feet."

There have been many suggestions and counter-suggestions as to the actual size of the orrery itself, and any conclusion I have come to has been based on an article in the Preston Guardian on the 12th March 1881, when it was sold following Annie Leonora's death, when it was described as 'somewhat ponderous,' which suggests it was a considerable size.

Moses completed his call to arms by asking if they were prepared to 'nibble at it?' He continued by saying, "The young men who had wives and sweethearts, should be anxious to be nibbling at it, and if they determined to be nibbling at it, they would before long have an institution that would belong to the working men of Preston," comments that raised even more cheers and applause.

The winter months that followed were packed with the necessary preparations for the bazaar, which was held over four days in the Exchange Rooms the Corn Exchange from Wednesday 25th April until the following Saturday. Admission charges for the opening day were one shilling (5p), followed by sixpence (2½p) on Thursday and Friday, and 3d on the final day.

The Institute reported that the appeal made by the Committee, particularly at the December Tea Party, had met with a response that was unprecedented, with the number and variety of articles received and promised that the event deserved to be regarded as an exhibition of both the arts and manufactures in its own right.

The bazaar was to include specimens of British and foreign workmanship, models of mechanism, architecture, and statuary. Specimens of various genres of natural history would be displayed, together with books and other publications which included autographed contributions from eminent authors. Various items, from scientific apparatus to flowers, fruits and other plants were also on display.

Finally, and probably most appropriately there was what was termed a 'Working Men's Stall', which comprised items that had been manufactured by the 'Operatives of Preston', purposely for the bazaar. These ranged from an extremely elegant and fashionable Victoria Pony Phæton, articles of furniture, clothing and *bijouterie,* or jewellery that is esteemed more for the delicacy of the workmanship rather than the value of the materials used to make it.

The bazaar was an immense success, as it needed to be. Originally, when the notion of a new building was suggested for the Institution, the immediate plan was for a building in Cross Street, adjacent to the long gone, but wonderful home of the Literary and Philosophic Society on the corner of that street and Winckley Square. The subsequent amendment to that idea, and to build on land on Avenham Walks in a more exposed position, would require the additional cost of two finished elevations rather than one, and a subsequent doubling of the overall cost from £2,500 to £5,000.

In the event, the building was completed by the following summer, although some of the interior work was left for completion when funds allowed. The main hall, which was capable of seating four hundred people, with a gallery that seated well over another one hundred, was completed, and put to immediate good use, as was the museum and library, the contents of which had now been extended to almost 4,500 volumes.

The first A.G.M. to be held in the building was the 21st in the Institution's existence, and the president, Thomas Birchall made reference to that by congratulating the members on attaining their legal majority and entering on their estate, 'by which' he explained, "I mean the handsome building in which I now have the honour, for the first time, of addressing a public assembly."

Suffice to say that the Institute had taken their place in a new home, a situation that was to continue until 1882 when arrangements were made to transfer the land, buildings and other property belonging to the Institute, to the trustees of the Harris Institute, which was being endowed by them with a sum of £40,000, for the purpose of teaching subjects in arts, science, literature, and technical and industrial education. The Institute for the Diffusion of Knowledge was dissolved on the 30th June 1882.

Other matters that occupied the time of Moses Holden during the rest of this current decade, included his Triennial Lectures in Preston, but they were the only such events. Despite the name, it had, in fact, been just over four years since he delivered his last. The announcement was accompanied by the information that during those years many new discoveries had been made, and several new planets added to the system, all of which it was his intention to describe and exhibit. The lectures were held during the last week of September, with a recapitulatory lecture being given on the 1st October, with reduced Box prices of one shilling and sixpence (7½p), and the

Pit and Gallery priced at one shilling. (5p) and sixpence (2½p) respectively. These were half the price of his previous week's lectures, and were intended for the benefit of the working classes, Sunday schools, and Sunday school teachers.

Moses was now in his 72nd year, but appeared to retain much of the rigour of his former years; in fact, during these lectures he took the opportunity to tell his audience that the passing years had only served to increase his ardour for his subject and his studies of it. He displayed all the most recent discoveries including the seven new planets, Flora, Iris, Métis, Hebe, Astraea, Hygeia, and Neptune.

Neptune had been observationally discovered in 1846, but only after it had been mathematically predicted by Urbain Le Verrier. Of local interest, the discovery of Neptune led to the discovery of its moon, Triton, by William Lassell, just seventeen days later. Lassell was, like Moses, a Boltonian; he was born in 1799 and lived for eighty one years. He was a beer brewer by trade, a livelihood that provided him with a fortune that allowed him to follow his passion of astronomy.

In 1815, and for a period of seven years after that, he was apprenticed to a merchant in Liverpool, and he later built an observatory at his home in West Derby, Liverpool. In 1815 his family had moved to Toxteth Park, the place you may recall where Moses erected the memorial to Jeremiah Horrox. They later moved elsewhere in the city.

Considering how much time that Moses spent in Liverpool, and their shared passion for astronomy, I would find it inconceivable that the two men were not acquainted, and yet I have never found any evidence.

His early days as a brewer coincided with the construction of the Albert Dock, and were a boom time for suppliers of the brewer's products. It was said that the amount of blasting that was carried out to create the docks, put the navvies at greater risk of death than the British soldiers who were battling at

Waterloo. In addition, they were each provided daily with up to twelve pints of beer whilst at work to quench their thirst, and to try and dissuade them from slipping away to the nearest alehouse!

Additionally, in the course of these lectures, Moses gave an account of the discoveries that had been made with the use of the 'far-famed' telescope of Lord Rosse. Lord Rosse was the Irish nobleman, William Parsons, the third Earl of Rosse, who had built his own massive reflecting telescope in 1845. The telescope, which had replaceable mirrors for when they became tarnished, was hung by chains from two fifty foot walls, with Rosse sitting on a platform fifty feet in the air, gazing down the tube. That sounds marginally more hazardous than life for a rock-blasting navvy in Liverpool?

Although I am sure that there must be a considerable amount of useful information tucked away in the pages of many local newspapers, particularly relating to Moses efforts as a Wesleyan preacher, not many seem to find their way into the pages of the main publications. Odd ones do, of course, such as the sermons that Moses gave at the end of May 1849 in the Association Methodist Chapel in Darwen, and again for the Wesleyan Methodists in Clayton Street, Blackburn, five or six weeks later in early July. Both were in support of the Sunday schools in the respective towns, and collections totalling eighty pounds were received; a huge contribution at a time when an average weekly wage was a fraction of that amount.

In 1850 Moses gave a sermon to the Wesleyans in Walton-le-Dale, and although the collection amounted to only two pounds and eighteen shillings. (£2.90p), it was said that it would defray the costs of six month's rent as well as a few other incidental expenses.

Chapter Nine

1851 – 1860

As is often the case with celebrity, and there's no doubt that in Preston, Moses had earned that status locally, demands on the individual's time can pose its own problems, but Moses, now in his 74th year seems to deal with those demands in a pragmatic and wholehearted way. This coming decade will see him present his Farewell Triennial Lecture course in Preston, and later a lecture to the Institute, but gone, it would seem is much of the travelling. An exception to that was in February 1851 he went to Ormskirk to present his course of astronomy lectures. There is no public record of these lectures, but I have come across a letter to Moses from James Dixon, the secretary of the Ormskirk Established Church Society, dated 5th March of that year, thanking him for his recent visit, and for his kind present of a copy of the Maps of the Visible Heavens.

Unlike his usual lectures, I think Moses either gave the course gratis, or for a fixed fee, for the letter recorded that the proceeds of the lectures far exceeded the expectations of the Committee, which made it sound as though the event was to raise funds for the church.

You will recall that Moses cared little for the name attached to a religion, although the Roman Catholic faith never seems to have benefitted from his sympathies. Indeed, before the end of the first month of this decade, he was invited to a social tea meeting in St. James's School, Knowsley Street in Preston. The purpose of the meeting was to form St. James's branch of the Preston Protestant Association, an action prompted by the attempts of the Rev. Dr. Gentili and others of the Roman Catholic faith to propagandise in the St. James's district. Moses was asked to address the meeting, which he did with a speech that was characterised as one that would encourage parishioners to rally round their minister.

The report of his speech began by describing its presentation as being delivered 'in his own felicitous manner,' which means 'strikingly apt, or pleasantly ingenious.' That description has been so applicable in many instances throughout his life – he always seemed to know the right thing to say at the right moment; an enviable talent was it not?

It would seem that a deal of solicitation had been taking place, with Protestants attempting to convert the Catholics, and the Romanists responding in like manner, for in his opening remarks he commented that 'the Pope saw the clergymen of the Church of England nibbling at Popery; and not only so, but he saw that Jabez Bunting had got hold of it, and had become Pope himself, or next door to it.' Moses enjoyed using the word 'nibbling' it would seem. Bunting (1779 – 1868), was a Wesleyan Methodist minister who completed the detachment of Methodism from its Anglican base; when he found Methodism it was a society, and he consolidated it into a church. He was following the Puseyism teachings of the learned Rev. Edward Bouverie Pusey, who was one of the leaders of what was called the Oxford Movement.

He told his audience that he had once been requested, by Mr. John Roby, author of *Traditions in Lancashire*, and who later drowned in the ill-fated *S.S. Orion* in 1850, to attend a Puseyism sermon in Rochdale. After attending the sermon, Roby had asked Moses what his opinion of the preacher was, so he had told him *'that his grandmother or mother had neglected to teach him the church catechism,'* a comment which caused great laughter.

Roby requested an explanation from Moses, and he replied, "The preacher made baptism mean regeneration. The catechism said baptism was an outward sign of an inward and spiritual grace." The audience applauded. He then added, 'I've no doubt Mr. Roby will have told the preacher what I said!' and the congregation laughed and cheered.

There were two eclipses during 1851, one of the Moons' immediately before the above tea party at St. James's and one of the Suns' towards the end of July. His letter to the *Preston Chronicle* predicting the January event read, *'the Moon will rise partially eclipsed at one minute after four o'clock, Greenwich Time (that time is used in Preston, or ought to be).'* One gets the impression that Moses was an 'ignorance is no excuse' advocate.

When he reported on the July eclipse of the sun, he reported for the first time on the drop in temperature during the period of the event. I'm sure that it's a routine procedure, but Moses never before mentioned it. On this occasion he was again at Howick House with Thomas Norris, and the thermometer they hung outside in the sunshine registered seventy eight degrees. After the eclipse began it fell gradually till the greatest obscuration when it stood at sixty six degrees. He commented also on the behaviour of the swallows flying round the house, 'making a great twitter and noise,' presumably confused about a particularly early roosting time. The darkest period of the eclipse occurred at 3.16p.m, and it had ended by 4.20p.m.

The 'Great Exhibition of the Works of Industry of all Nations,' or more simply, 'The Great Exhibition,' and sometimes referred to as the 'Crystal Palace Exhibition,' was held between 1st May and 11th October, 1851, in Hyde Park, London.

Both the Preston Chronicle and Liverpool Mercury in August 1851 carried the news that 'As a mark of respect, a few of the friends of Moses Holden, have this week presented that gentleman with a purse containing upwards of £22, to defray the expenses of a visit to the Exhibition.

Whether his visit was self-inspired by his interest in mechanical construction, or whether the visit was inspired by others, isn't known. Neither is there a record of whether he attended the event with anybody else, but at least one local tradesman was to benefit from the exhibition. Thomas Yates, watchmaker, trading from 159 Friargate in Preston, later advertised that he had been awarded a prize from the Great Exhibition, as well as a large silver medal and honorary testimonial from the London Society of Arts.

A double page advertisement for Yates' wares in Slater's Trade Directory of 1855, carried not only this information, but also a testimonial from Moses Holden, dated 1847, that read, *'Having one of your Patent Watches, with a fine chronometer balance, I have tested it for some time, and find that it is capable of the finest adjustment, and, when properly rated and regulated, will keep time equal to one of the best pocket chronometers. And, I have no doubt, will, in the end, supersede that instrument, because the spring and chain will not be so liable to break, as there is not half the power required in yours that there is in the pocket chronometer.'*

A striking similarity to the first few days of 1852 compared with the previous year saw another eclipse of the Moon, although this time it was to be a total and visible eclipse, with the Moon entering the Earth's shadow at 4.21am on the 7th January, with the greatest darkness beginning at 5.21am precisely. Moses predicted that after that time, and for a period of almost an hour and forty minutes, the Moon would appear copper-coloured.

As is often the case when an interest requires fair play from the weather, it failed to oblige. When Moses rose to witness the event he described it as very dreary, wet and cloudy. Although a good deal of light was being given out from the clouds, by the full moon being behind them. "I kept a good look out, but no moon appeared, nor did I see any opening in the clouds at all." It wasn't until seven o'clock that the clouds began to break, and within ten minutes the planet Mars appeared. "I took the telescope, and a power of 144, and saw its polar snows beautifully; the air was so very clear after the rain." Moses concluded the report by adding that "at fifteen minutes past seven the clouds cleared off the Moon, and I saw that two-thirds of it was still dark; I then observed the eclipse to the end, just before eight o'clock."

It is difficult to know how much interest Moses created through his comprehensive and descriptive forecasts and accounts of astronomical events, or how Preston compared with other towns and cities. It has to be assumed that ownership of even fundamental optical equipment will have been restricted to the manufacturers in the growing economy of the town, and to the upper middle classes and the professionals.

Moses often spoke of allowing people to view the heavens through his telescope at his home in Jordan Street, but there is no evidence to support the idea that the operative classes were welcomed. I'm sure they wouldn't have been turned away, but it's disappointing that there's no positive evidence.

During the time between the eclipse and his Farewell Lectures in October, the only noteworthy event was the appearance of the planet Venus during the early days of May. He wrote:

THE PLANET VENUS

Venus is now a fine object for the telescope; its appearance is like the moon at the first quarter, or half enlightened; and being near its greatest north declination, it is high up in the heavens, and will not set until late at night on the 8th May, at 11.52pm. Its greatest brilliancy will be on the 14th June. Then it will be a crescent, and will continue to enlarge in diameter, and the Horns of the crescent to grow finer and sharper till its inferior conjunction with the sun, which will take place on the 20th July, next.

Jordan Street.   May 5th 1852

M. Holden

Moses official retirement from his tour of lectures was a highly anticipated eleventh and final presentation of his Triennial Lectures. They were advertised three weeks in advance, and an article that appeared in the *Preston Chronicle* just two days before the first lecture read, *'The subscription list to Mr. Holden's farewell course of lectures on Astronomy, is a more numerous one than he had hitherto on any occasion been honoured with. We are pleased to see that our old townsman, on retiring from his pursuit of lecturing, is receiving such a mark of esteem from among those whom he has so long lived.'*

The three parts of the course were attended to capacity, and the accounts of receipts and expenditure can be seen in Appendix Eight. There are no detailed accounts, however, for the recapitulatory lecture that he gave on the 18th October at half the usual admission fees, other than the total receipt on the night. A report of the recapitulatory lecture revealed that in addition to the use of Moses *'splendid orrery and a great number of telescopic views,'* it also mentioned that he *'demonstrated how the heavenly bodies performed their different revolutions round the great central attraction – the sun – which was strikingly shown by a piece of ingeniously contrived machinery representing the whole of the solar system.'* That is the function of an orrery, but the article tends to make it sound as though this piece of equipment was different from it.

A list of the sponsors for the Farewell Lectures was particularly notable for the number of names that also appeared on a list of the attendees at a banquet for the Mayor at the Bull and Royal Hotel earlier in the same year; a veritable who's who of the gentry of Preston in the middle of the nineteenth century.

It is interesting to note that when Moses' daughter died in 1881, a subsequent auction of furniture and items of *'memorabilia belong to Moses Holden,'* included his orrery and organ, both of which were used during the course of his lectures. Indeed, it has been said that Moses played the organ himself during the lectures, presumably to give atmosphere to the proceedings. Perhaps his audiences and his commentators regarded it as an integral part of the whole and believed that it required no special mention as a matter of course.

FAREWELL LECTURES.

THEATRE ROYAL, PRESTON.

OURANOLOGIA,

OR THE HEAVENS DISPLAYED,

IN A

Course of Astronomical Lectures,

ILLUSTRATED WITH A MOST BEAUTIFUL GRASTRODIAPHANIC OR

GRAND TRANSPARENT ORRERY,

TWENTY FOUR FEET IN DIAMETER, WITH SUPERB SCENERY;

The Sun, Moon, Planets, and Stars,

SHINING AS THEY DO IN NATURE, ENLIGHTENING ALL THE PLACE.

MR. M. HOLDEN,

PRACTICAL ASTRONOMER.

Returns his grateful thanks to the Ladies and Gentlemen of Preston and its vicinity, for the Patronage bestowed upon him on former occasions; and, after a lapse of three years, again presents himself before them, and most respectfully informs them, that he

WILL DELIVER, FOR THE LAST TIME,

HIS COURSE OF THREE LECTURES,

With all the recent Discoveries, familiarising the sublime Science of Astronomy, by first describing, and then exhibiting the same on his Splendid Machine, illustrating the grand Operations of Nature, as displayed in the Motions of the Heavenly Bodies, with all the Phenomena arising therefrom, such as Day and Night—the Rising and Setting of the Celestial Orbs—the vicissitudes of the Seasons—Full and Change of the Moon—Eclipses—Transits—Comets—Constellations—Nebulæ, &c.

With an Immense Quantity of Splendid Scenery,

Consisting of Telescopic Views, &c., surpassing any thing of the kind ever exhibited in this kingdom for correctness; drawn by the Lecturer, from Views by Telescopes of great power, made by himself, which Drawings he will have the pleasure to exhibit in this Course, to a Scientific and Discerning Public, whose Patronage he would humbly claim. The Lectures to be delivered

ON MONDAY, 11th, TUESDAY, 12th, AND THURSDAY, 14th OF OCTOBER, 1852.

SUBSCRIPTION:

FOR THREE NIGHTS, BOXES, 8s. PIT, 5s. 6D.
(A COURSE CONTAINS THREE TICKETS, ONE FOR EACH EVENING)
SINGLE NIGHT, BOXES, 3s. PIT 2s. GALLERY, 1s.
TICKETS to be had of Messrs. Dobson & Son, Mr. Clarke, Messrs Addison, & Mr. Oakey, Booksellers; & of the Lecturer, No. 14, Jordan Street, where Family Tickets may be had.

DOORS OPENED at HALF PAST 6, & the LECTURE to COMMENCE at 7, EACH EVENING.

M. H. has published a Fourth Edition of his beautiful concise ATLAS OF THE HEAVENS, in 23 Maps, 12mo., whereby all the Stars, visible in the Latitudes of Britain, may be easily known, with their names. The work may be had at his residence, and at the Lecture. Price 3s. 6d.

An item of particular note was a separate slip of paper that accompanied the accounts for the lectures. Precise use of the first item isn't clear, but the others are perfumes, and one can only assume that theatres were generally damp, and consequently didn't smell as good as they could. A combination of the various aromas may have counter-balanced that distinctive and unwelcome aroma:

Spirit Wines1 quart Oil of Lavender1 oz
Essence of Rose½ oz. Oil of Neroli10 drops
Atta Rose15 drops
Essence of Ambergris ¼ oz
Essence of Mille fleur ½ oz
Essence of Jasmine   ½ oz

In September 1852, William Rogerson of the Greenwich Observatory wrote a letter to Moses, which he had published in the *Preston Chronicle.* It was an explanation of his inability to, and sadness at not being able to attend his final lectures. He had attended them previously in Preston as well as elsewhere, and had often expressed a desire to witness them once more.

His letter was full of eulogistically expressed thoughts that made apparent how important their friendship was to him. *'Last night,'* he wrote, *'while I was surveying the spangled robe of night, I thought of you; I fancied I heard you in your usual energy, in the crowded theatre, pointing out the wondrous works of the Creator.'*

He closed his letter by saying, *'Nine and twenty years have passed away since I first had the pleasure of seeing you in this vale below. Similarity of mind united us; and since then we have many a time conversed together, and when distance has prevented, have corresponded with each other on those subjects on which we delighted to meditate. I trust that when this short life is over, we may, through virtue of the Redeemer's merits, be admitted into the celestial regions, to spend an eternity together in intellectual enjoyments.'* Within six months, Rogerson was dead, and I feel certain that he knew his days were numbered when he wrote the letter.

Shortly after Rogerson's death, and unrelated to it, Moses became involved with an attempt to revive the Preston Samaritan Society. In late May 1853 Moses preached a sermon in the Wesleyan Methodist Chapel in Chadwick's Orchard on the Sunday evening in connection with the Society. Its later address was Liverpool Street, a thoroughfare that ran along the northern extremity of the Covered Market. The following evening a public meeting took place in the Exchange Rooms of the Corn Exchange, where the chair was taken by the Rev. T. Clarke of Christ Church. Speeches were made by him, Moses, and a number of other men, and a number of resolutions were agreed upon, but although it was said that 926 people had been relieved during the past twelve months, the funds of the Society were in a perilous state.

The following year, Moses had determined that he wouldn't attend the annual meeting because of a bad cold, but he had been requested by so many people to do so that he had relented. When he stood to address the meeting he was received with loud applause. In his speech he requested his audience "to think of the poor this cold and very severe weather. The atmosphere", he continued, "had not been so cold in Preston since the day on which Nicholas Grimshaw died, when the thermometer stood one degree lower than this winter. It was only the poor who could tell the severity of the weather; they could tell it because they felt it. They might go into many a poor man's house this severe weather, where they could find him, perhaps, sick in bed; the children crying for bread, their mother having none for them; the fire may be out, the windows let in the cold and rain, and many other things beside, which those who sat round their snug fires never thought of." He went on to say, "What a mercy that such a society as this existed. The visitors of the society went into such houses as I have just described, many a time, with a shilling in their hands. But what could a shilling do for a man, his wife, and family? It had done wonders; it had opened the heart of the receivers; it had enabled the visitors to pray with them, when they were not at first willing; nay, it had learned them to pray themselves."

Moses' age seems to have been catching up with him, for it would appear that he was an absentee for all the Society's events in 1860 and 1861, but he spoke briefly at the February 1862 meeting when he said that although he was now feeble and old, for he was in his eighty fifth year, his affection for the Society continued. The Samaritan's Society had been founded in December 1816, and Moses had had an involvement since at least 1845.

Only two events of astronomical note occurred in 1853, a Comet that was present from the end of August through to the first few days of September. Moses had observed it through an equatorial telescope, although he explained its position and said that it was visible even to the naked eye. In a letter to the *Preston Chronicle* Moses stated, *'I judged it to be the great comet of 1556, because its track in the heavens very nearly agrees with that assigned by Mr. Hind, for August and September, in his book the "Great Comet" of 1848, pages 64 and 65; but he (Hind) states in the Times, that it is not that comet.'*

John Russell Hind was a native of Nottingham, and a few days later an article appeared in the *Nottinghamshire Guardian* which read, *'We were much amused to observe in our Preston contemporaries the somewhat precipitate announcement that Mr. Moses Holden, had discovered the comet to be identical with Hind's. Very strange all this: when Professor Hind himself repudiates the idea, and maintains that the avatar (or return) of his comet still to be in the future.'*

There was less controversy about the sighting of an Aurora borealis, described as grand and very bright, which crossed the heavens at right angles to the magnetic meridian, reaching from the eastern to the western horizon. He recorded that he had seen more than twenty such arches, but never one that was so brilliant, nor so strangely bent. This one was visible from 9p.m on the 9th September, until after midnight.

Moses' attendance at the 24th meeting of the British Association for the Advancement of Science, which was held in the September of 1854 at St. George's Hall in Liverpool, demonstrates the sort of company with whom he felt at ease. Among the seventeen-man committee were characters like Abraham Follett Osler, the meteorologist and chronologist, who played a significant role in the synchronisation of church clocks with Greenwich Mean Time at a time when the railway timetables were starting to play a significant role in the economy of the country. Jean Bernard Léon Foucault, the French physicist who is best known for the Foucault Pendulum, a device that demonstrates the effect of the Earth's rotation, and an excellent example of which there is in the ground floor rotunda of the Harris Museum and Art Gallery at Preston.

Another member of the committee, and a man who had worked as an active member and official of the British Association for the Advancement of Science from soon after its inception in 1831, was the Rev. Dr. William Scoresby, who was, like his similarly named father, an arctic explorer. The Association was a breakaway movement from the Royal Society, or more fully, the Royal Society of London for Improving Natural Knowledge, a society that had a history back to 1660, and was seen by some to be elitist and ultra-conservative.

Another distinguished member of the committee was Captain, later Admiral, William Henry Smyth, an English naval officer, who was also an accomplished hydrographer and astronomer.

During the first full week of March 1855, Moses gave a lecture in the theatre of the Institute. The title of the lecture was 'Meteors, commonly called shooting stars'. Unfortunately, the newspaper report that followed the event was totally inaccurate, necessitating a letter from Moses to make amends. It is one of the few occasions where we see the content of one of his lectures, which arose from the paper's claim that he had said, *'meteors are the fragments of planets which had been shattered by comets in their eccentric course through the heavens.'*

Moses wrote in the following week's newspaper, *'The fact is I never had such a theory, nor did I once mention a comet in my lecture that evening. I said the continental astronomers believed that a planet moving betwixt Mars and Jupiter had been burst, and the thirty-one small planets moving there (where there should be but one) are the fragments of that body. I also said that a vast avalanche of matter would be launched into space, with millions of small bits, and these are revolving round the sun, but would not have sufficient projectile to maintain their orbit, therefore they had passed Mars, and in their revolutions were passing the earth; and the last time this vast quantity of matter approached the earth was on the 12th and 13th November, 1833, when the grand display of meteors was seen in America and elsewhere.'*

He went on to explain that 'When one of the large fragments or stones enters the earth's atmosphere, the velocity is so great, and the resistance so powerful, that the stone is red-hot, and is noticed as a fireball, and bursts with a tremendous noise, and a shower of meteoric stones fall to the earth.'

The total eclipse of the Moon on the 2nd May 1855 was spectacular for those who witnessed it at its peak at 3.16a.m, compared with the partial eclipse of the same body on the previous 4th November. On that occasion it occurred at the more reasonable time of 9.12p.m, but had only obscured the Moon by half a digit, or one twenty-fourth of its diameter. One for the astronomers, I would suggest, rather than the interested onlooker?

I mentioned earlier that local celebrities appear to have a lot of demands made upon their time, and an interesting example was a meeting in the Theatre Royal on the 1st October 1855. The meeting had been called by the promoters of the Preston Early Closing Association, whose aim was to provide an additional few hours leisure for particularly the retail shop employees. Why this should be of interest to Moses isn't immediately clear, but the meeting received the unanimous backing of all who attended it.

The following week, an advertisement appeared in the *Preston Chronicle* which listed all the establishments who had determined to adopt the proposals. Those proposals were to see the closing of retail outlets at 7p.m (with the exception of Saturday evening), between October and the end of March, and 8p.m from April to the end of September.

When were these working people expected to take advantage of the benefits offered by the Institute for the Diffusion of Knowledge? I'm sure it was a question that Moses would have liked an answer to, and was certainly one of the reasons for his interest.

A closer link to Moses' interests is to be found in a series of fund-raising lectures he gave around this period in connection with the Mutual Improvement Society for the town's churches. In December of 1855 he gave a lecture on astronomy for the benefit of St. Mary's Church, Ribbleton, and in April 1856 he gave the final lecture in a series of three which were given for the benefit of Christ Church in Jordan Street. In early November 1857 he gave a further lecture in aid of Christ Church, with the particular improvement in the latter two instances being the provision of the clock for the tower that was discussed when the church was built in 1837.

Correspondence

Preston Chronicle 15th December 1855

THE PLANET SATURN is now a fine object every evening. About seven o'clock it is in the east, moving towards the south, and is the largest star in that part of the heavens. At midnight it is directly south. Through a good telescope it is singularly beautiful, the ring being this day at its greatest opening, and the planet being at it's nearest the earth for this season. It will remain a good telescopic object for a year or two, although it will gradually lessen in the extent of its opening.

It will be the spring of 1885 before the ring will be again seen as it now is. Our venerable townsman, Mr. Moses Holden, having favoured us with a drawing of the planet as it appears through a telescope (not an inverting one), we have engraved it for the benefit of our readers. We may add that in the year 1863 the appearance of the ring will be as a straight line across the planet, and in 1870 as the subjoined cut inverted would represent:-

---

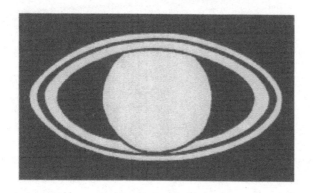

Saturn

The only things that have come to my attention from 1856 are two letters that Moses wrote to the *Preston Chronicle*. There had been an article in the *Chronicle* the previous week, informing the public of the occurrence of an eclipse, and noting that it would be an almost total eclipse, with only a small portion of the upper part of its body visible. His two letters appeared in the same issue of the paper, with the second one being supplementary to the first. It would seem implausible for this to be the only comment-worthy astronomic event of the year, but it may have been, for in other respects Moses still seemed to have plenty of enthusiasm in store.

Correspondence
The Eclipse of the Moon.

On Monday evening last the air was remarkably clear and fine, the stars shone with a lustre that we have not seen for some weeks, nor did we expect to have so clear a sky at the time of an eclipse, considering how cloudy it has been of late.

I saw the eclipse commence at the time you published in your last week's paper, 9hr 21mins 6 secs, and the middle was, as near as I could judge, at 10hr 54mins., but the end seemed to linger a minute or two later than 12hr 27mins 6 secs., so that it ended between 28 and 29 mins after 12 o'clock.

The moon looked like a polished ball of copper; the ruddy appearance was seen after the moon had passed halfway into the shadow, and continued till it had got more than half out of the earth's shade. The large star that appeared on the west of the moon was the beautiful planet Jupiter. I saw it through the telescope about midnight; its four moons and belts appeared grand.

15th October 1856

Ruddy Appearance of the Moon in Eclipses.

Your readers are aware that lunar eclipses are caused by the moon passing through the earth's conical shadow, or by the shadow of the earth falling upon the moon; and that they are visible in all parts of the world which have the moon above their horizon; and are everywhere of the same magnitude, with the same beginning and end. If the earth had no atmosphere, the moon would be as invisible in total eclipses as she is when new.

The cause of her being visible, and of the copper-coloured glow, is thus explained:- The rays of the sun on entering our atmosphere are bent inwards from their original course; and they again undergo an equal deviation, in the same direction, when leaving it; the amount of the first refraction being thus *doubled,* the rays are enabled to enter the earth's shadow and fall upon the moon: the invisible vapour in the lower strata of our atmosphere imparting to them the ruddy hue of sunset.

Should, however, a belt of clouds from 30 to 60 degrees in breadth surround that great circle of the earth, in which at the moment the sun is seen in the horizon, or 90 degrees from where the moon is vertical, little or no refracted light may reach the moon; and she may for the time be completely obliterated, as in the eclipse of June 5th 1620, and April 25th 1642.

16th October 1856
Jordan Street, Preston
Moses Holden

The opening month of 1857 saw an occultation of the planet Jupiter by the moon. Moses wrote to the Preston Chronicle with his report of it:-

On Friday evening, the 2nd January, the moon covered the planet Jupiter. I saw the occultation take place, with a power on the telescope that made Jupiter look as large as the moon does to the naked eye. All the four moons were visible, but clouds came on, and the next opening in them showed that the fourth moon, which was at its greatest distance, was covered; and in another opening I saw the dark edge of the moon come on to the edge of Jupiter, and soon cover the planet. This took place at 5hrs and 2mins.

I saw the planet come from behind the bright edge of the moon at 5hrs 54mins.

The appearance of Jupiter was very beautiful, and round, even to the dark edge of the moon; and the same when it came out from the bright edge; it was not affected by any refraction whatever.

9th January 1857
Jordan Street
Moses Holden

The Rev. Robert Brickel and the Church of St. Michael's, Hoole, Near Preston.

The fact emerged during 1857 that the Rev. Robert Brickel, the then incumbent of St. Michael's, was anxious that a memorial ought to be created at the church in connection with the distant presence of Jeremiah Horrocks at that church in the seventeenth century, and the reason for his notoriety. As Mr. Thomas Turner Wilkinson, F.R.A.S., a headmaster in Burnley, and member of Burnley Mechanic's Institute worded it, 'The Rev. Brickell's endeavours will identify Horrocks with the parish until the end of time.'

I am including this story because back in 1826 when Moses provided for the memorial to Jeremiah Horrocks in Toxteth Park, he wrote, as I mentioned in chapter five to the editor of the Preston Chronicle, Isaac Wilcockson, informing him of his intention. In the letter he made clear that if he didn't get the necessary permission to have it placed in St. Michael's, Toxteth Park, he would site it in St. Michael's, Hoole.

The Rev. Brickel, who had arrived in Hoole in 1848, made a nationwide appeal to astronomers and others who took an interest in scientific pursuits, initially to engender interest in his notion. It wasn't clear what form the memorial would take, but a clock on the old tower had been a suggestion, to accompany the prominent sun dial on the same tower. It was said that there was reason to believe that Horrocks was the man responsible for the placing of that sun dial, but that has to be erroneous because the tower hadn't been built when Horrocks was there.

The article in the *Preston Chronicle* went on to comment on Moses Holden's memorial to Horrocks in Toxteth Park, and finished by observing that they thought that Brickel's idea would meet with general approval.

At the end of July, the Rev. Brickel wrote from his home at Manor Bank, or Manor House, in Hoole, to the *Preston Chronicle.* In it he gave readers an update on his efforts, which included writing to both the Liverpool and Manchester newspapers, and expanded on what had been said in the original article by saying that he would prefer "a handsome monument or a stained glass window." It was only when a decision had been taken as to the nature of the memorial that the amount of money needed to provide it would be known.

In a letter a couple of weeks later, Brickel again raised the subject of money, mentioning a collection of between two and three hundred pounds, with which he proposed the provision of a 'Horrocks Free Parish Library' as well as another monument.

He also listed the extraordinary achievements of Horrocks, despite the fact that he was dead by the age of twenty two:

a. First to predict and then observe the transit of Venus,

b. First to reduce the sun's parallax near to what it has since been determined,

c. First to suggest the correct theory of lunar motions,

d. First to remark on a phenomenon proving the extreme smallness of the apparent diameter of the stars,

e. First person who began a course of tidal observations, and

f. First to devise the beautiful experiment of the circular pendulum for illustrating the action of a central force.

In the east window of St. Michael's is a depiction of
Horrocks observing the Transit.

By mid-September Brickel was able to report that the Lord Lieutenant of Lancashire, the Mayors of Liverpool, Manchester and Preston, the leading scientific and influential gentry of the neighbourhood, as patrons, committee, and unsolicited subscribers, had guaranteed that his endeavour would be supported. The ongoing cotton famine meant that the timing of the appeal rendered it unfavourable, but Brickel pressed on.

A week later another letter appeared in the Chronicle, with the first line reading, 'Beggars are generally deemed troublesome,' so it was clear what was to follow. Now that the Lord Lieutenant of the County had an involvement, Brickel deemed it a county matter, and wrote to the entire county M.P.'s, followed closely by the county magistrates. However, it was the Preston Corporation on whom he focussed his efforts, initially on the basis that Preston was the closest town to Hoole, but also because the treasurer of the appeal committee was the current Mayor of Preston, and he thought Preston should set an example for other towns to follow.

In August 1858, Brickel wrote a further letter to the *Chronicle* with the heading 'The Horrocks Memorial', and a sub-heading 'Horrocks, the pride and boast of British astronomy,' a quote from Sir J. Herschell. In the letter he made reference to the fact that the late M. D. Whatton of Manchester had amalgamated a huge amount of material with the intention of writing about the life of Horrocks. Unfortunately Whatton died in 1830, but at the present time it was known that Whatton's son, the Rev. Arundell Blount Whatton, was preparing to complete his father's work.

In his letter, Brickel is making an appeal for genealogical information about Jeremiah, 'for there are strong reasons for supposing that he was from the same family as the late John Horrocks', M.P. for Preston, and early cotton manufacturer in the town. Whether there is a family link between the two men is unknown, so perhaps it was a useful marketing ploy on behalf of the Reverend.

Whatever it may have been, a further letter to the Chronicle in November 1858, revealed that the result of Brickel's efforts were more than satisfactory, with support from the head of the House of Stanley, the Lord Bishop of the Diocese, the learned master of Trinity College, Cambridge, the late and present Lord Lieutenants of the county, the mayors and a few county magistrates, and some county families. He reported that funds were currently being expended in a Horrocks Chapel, a memorial window, and a plain marble tablet. Twelve months later, in November 1859, Brickel reported to the readers of the *Chronicle* that the work had been thoroughly carried out, including a chancel-aisle, three windows and a tablet, plus a new clock which had been provided by the parishioners of Hoole alone. The memorial chapel was said to contain 'thirty sittings free to the poor forever.'

Although Moses Holden was rarely mentioned during the course of completing the work made possible by the fundraising efforts of the Rev. Brickel, he was a member of the committee formed to help achieve their aim. His apparent physical absence from the day-to-day dealings of the committee is probably an indication of his failing health and consequent inability to get more closely and actively involved.

The Great Eclipse of the Sun: Monday 15th March 1858

Regardless of any sign of failing health, Moses was certainly prepared for a once in a lifetime event, labelled the 'Great Solar Eclipse,' and sub-labelled, 'the most interesting solar phenomenon of the present century.' In a letter to the *Preston Chronicle* Moses wrote, 'I have had it calculated and drawn out for thirty years, though I did not expect to live to see it.' It was with more than a touch of irony that despite living long enough, he didn't see it!

Moses went to great trouble to describe how to make a dark glass through which the eclipse could be safely observed, 'using two pieces of equal and oblong pieces of glass of a convenient size for a waistcoat pocket.' The two pieces of glass were to be warmed in front of a fire before being smoked over a candle, the density of smoke varying from one end to the other. The two pieces were to be separated from each other by a small piece of card and taped together, with the varying densities of smoke allowing safe usage however bright or dim the sun may be.

Moses forecasted that the total eclipse would last just twelve or thirteen seconds, and that during that period the annular ring or circle would be only just discernible, and he joined his friend Thomas Norris at Howick Hall to witness the event.

Throughout the country, unfortunately, the eclipse itself was eclipsed; in some places totally, and in most others during a good portion of the obscuration. The Sun was not visible at all in many places 'on account of the thick, dark clouds which formed an impenetrable barrier to the vision.'

The report went on to explain that we were more fortunate in Preston. 'There was a view, more or less uninterrupted, obtained of the eclipse, for a quarter of an hour or twenty minutes, and this fortunately comprised the time of the greatest obscuration.'

Disappointingly, the reports of the event that the Chronicle had been hoping for were unforthcoming, for as indicated earlier, Moses had gone to Howick Hall where he expected visibility to be better, but unfortunately, the temporary break in the clouds enjoyed by observers less than two miles distant, did not occur at Howick, and not a glimpse was had by Moses or his Howick friends of either Sun or Moon during the entire period.

***

The disappointment of missing the eclipse was perhaps reduced a little by the appearance, in September the same year, of a magnificent comet, that had been discovered on June 2nd by Dr. Giovan Battista Donati, and now bears his name. By early September it was said to be visible to the naked eye. In an article in the *Preston Guardian* of the 4th September, Moses Holden said that, "On Sunday 5th September, the comet, the Pole star, and the Pointers, will be in a straight line, the uppermost of the Pointers being in the centre between the two. The position of the comet in the heavens may therefore be readily ascertained. The Pointers are two of the bright stars in Ursa Major, more familiarly known as "Charles's Wain" (the Waggon and Horses), or "The Plough," which it more resembles, and is so called because if a line is drawn through them, it will direct the eye to Stellar Polaris, or the Pole Star. In the present instance, if the line be carried downwards, the same distance, the exact position of, and probably the comet itself, may be detected."

A month later, at the beginning of October, Moses reported having measured the length of the tail of Donati's comet, and found it to be an incredible thirty degrees. It is perhaps worth knowing that although a comet acquires the name of the discoverer, it is by no means certain that it will be retained. The rule at the time of this comet was, *'that he who is the first to determine the orbit of a new comet, bring to light its past history, and successfully predict the period of its return, has the honour of handing his name down to posterity.'*

In an interesting letter written by the distinguished astronomer we met earlier, John Russell Hind of Nottingham, to the *Times*, dated 1st September 1858, details of the comet's forecasted progress were given. He then made some comments that, in part seemed strange, but also of interest as far as the 1811 comet was concerned. You will recall that Moses presented a paper to the Preston Literary and Philosophical Society in 1812 concerning that comet. Hind's comments read, *'It is evident that the inhabitants of Venus will see the celestial visitant to great advantage about the middle of October, and probably will retain as lasting a remembrance of the comet of 1858 as we terrestrials do of the famous comet of 1811.'*

At the age of seventy nine years, we find Moses still fundraising on behalf of the Christ Church Mutual Improvement Society. On this occasion he gave a lecture on telescopes in mid-November in the National School-room in Bow Lane. In a well-filled room, Moses sketched the history of the telescope from the time of Galileo down to the present day. It was said that several of his anecdotes were very amusing, and will be long remembered. Once again, isn't it a pity that no examples are available for us to share?

He was also still attending the annual meetings of the Preston Samaritan Society, and still happy to address those gathered in the Lune Street Chapel schoolroom in Fox Street, as he did in the November of 1858. Whilst the work offered by the Society was still a necessity, and the relief that was being effected by its work still welcomed, its finances continued to be little different than they had been for a long time. It was said that during his long life he only ever missed two Wesleyan Missionary breakfasts, and only four meetings of the Samaritan and Bible Study groups, and he only missed those *'as a consequence of old age, and a dread of the cold draughts on the stage of the Theatre'* where many of them were held. Every effort was made to engage with the parts of Preston's society who were able to make a significant difference to the needy, but it must have been particularly frustrating when a concurrent appeal to raise money for a memorial to the late Prince Consort had raised in excess of £400. Their beloved Queen may not have been amused at such disparity.

After Moses had delivered an impressive speech at this meeting, he moved *'that the thanks of this meeting be tendered to the committee and officers of this society, for their past* services, and that they be respectfully requested to continue them.' The Rev. R. Maxwell, in response to Moses, said that "he hoped that their venerated friend Mr. Holden's days might be increasingly happy, and that his last day would be the happiest of all." I'm sure the comment sounded better than it looks in writing! In the event, he lived another five and a half years.

On the afternoon of Christmas Day 1858, Moses addressed a function held by the Methodist Free School in Chadwick's Orchard. Around 300 people, including 240 of the junior scholars enjoyed a Christmas Tea-party, which was provided free of charge by Robert and John Hoyle, and John S. Orford, all local businessmen. After tea, the president of the school, the Rev. John Guttridge addressed the gathering, tracing the origin and progress of the Sunday school institution. Guttridge Memorial Church in Deepdale Road, Preston, recently closed, was later named after this gentleman.

Just five days later, Moses attended another tea party, this time the 17th annual meeting of teachers, scholars and friends of the Christ Church Sunday Schools. It was held in the girls' new school, Wellfield Road, and more than 450 were in attendance. Many more were denied admission because of inadequate space in the building. After the business elements of the meeting were completed, Moses and several ministers delivered earnest and suitable addresses. The church choir, with their organist, Mr. Shaw playing the harmonium for the occasion, were in attendance, and led the singing of hymns, anthems and recitations, at intervals throughout the evening.

As far as the Methodist churches and chapels were concerned, Moses lived long enough to see those that he had helped to organise, in a vigorous and flourishing state.

Chapter Ten

1861 onwards

A distinct curtailment of involvement with all the many and varied facets of Moses' life had been evident for a year or two by the beginning of the 1860s, and unsurprisingly it continues that way until his death in 1864. Although there were a few exceptions, it was said that he was rarely seen outside Jordan Street during his last five years. Like many people of that age, he had good days and bad, and he was more or less confined to his house from the Christmas of 1863 for around six months, although it was only his last two days where he found himself confined to his bed. He was said to be in full possession of his faculties to the end.

A comet that appeared around the end of April 1861 brought a letter to the *Chronicle* from Moses, but it lacked the enthusiasm and detail with which its forerunners had oozed. His letter, written on the 10th May, even carried a comment that he – Moses – was surprised that Mr. Hind of comet notoriety, 'had not given us anything on the subject.'

This lack-lustre letter would appear to be the final letter written by Moses to the local paper, and for a man who was at the forefront of everything with which he involved himself, when the end came, matters were conducted with the utmost privacy. It isn't known whether it was an expressed wish of Moses himself, or of his wife, but considering his wide range of associates in science, education, the church, and the town in general, it is surprising that greater numbers weren't present at his farewell.

The bells of Christ Church slowly rang out, deeply muffled, on the night that Moses died, Friday 3rd June 1864. They rang out again as the cortege moved off from Jordan Street at the arranged time of 10.30a.m just one week later.

The cortege consisted of a hearse and three mourning coaches, plus the private carriage of Thomas Norris Esq., of Howick Hall. Of his immediate family, there was his wife, Isabella, his daughter Annie Leonora, and two of his grand-children, Susan Isabella, a child of William Archimedes, and Master T. Holden, who is now known to be Thomas, the grandson of Moses' elder brother, John.

The first mourning coach carried Isabella, Annie and the two children, the second carried the Rev. Canon Parr, Mr. T. Norris of Howick Hall, Mr. William Dobson, and Captain J. Duke of East View, said to have been an old acquaintance of Moses. The third coach carried Mr. R. Porter of West Cliff, Mr. Hall of Theatre Street, and Mr. Meek. It isn't known who occupied Mr. Norris's private carriage, for the blinds were drawn.

On its arrival at the door of the Church of England Chapel, a group of around fifty people had gathered to await Moses remains, and to pay their last respects. The service was conducted by the Rev. Canon Parr, and the funeral was conducted by Messrs. Meek Brothers, the undertakers of Preston, and for which they had provided one of their 'patent air-tight metallic coffins.' He was carried from the Chapel to his final resting place, and interred in a newly-dug grave, sixteen feet in depth just a few dozen yards from the Miller Road entrance.

The headstone in the centre, mid-distance with the trefoil aperture and pale weathering streak down the centre, is the resting place for Moses, his wife Isabella, and daughter Annie Leanora. Whilst William Archimedes' death is recorded on the stone, his body was never brought back from Melbourne following his death in 1852.

All the Preston newspapers, the *Chronicle, Guardian,* and *Pilot,* carried lengthy obituaries the day following his death, and on the 11th June, the day following his funeral, the *Guardian* contained a synopsis of his life that ran to twenty column inches, itself a clear indication of the esteem in which he was held in the town.

Following Moses' death, a meeting of the Institute for the Diffusion of Knowledge that he had helped to found was called, as a result of which, R. Newsham, the chairman, sent to Mrs. Holden a letter of condolence on her bereavement, and enclosed with the letter the details of two resolutions that had been passed at the meeting. Firstly, *'that the committee desires to record its high appreciation of the talents and laborious studies of the late Mr. Moses Holden, in the science of astronomy, and to express its deep regret at his removal from this earthly sphere,'* and secondly, *'that a copy of the foregoing resolution be sent to Mrs. Holden, his widow, with the sincere condolences of the committee; and that Mrs. Holden be considered entitled to all the privileges of a Life Member of the Institute.'*

Thomas Yates and the Parcel of Sovereigns.

In early 1865, Thomas Yates, the watchmaker of Friargate, bought a regulatory clock which had been the property of Moses Holden. When he looked inside the casing he discovered a small, carefully tied little parcel containing eight sovereigns in mint condition. He immediately returned the money to a very grateful Mrs. Holden who was most pleased to receive the windfall. She concluded that it may have been there for as long as thirteen years, and that Moses had probably forgotten about it.

Ten years after his death, an article appeared in the November 1874 edition of Longworth's *Preston Advertiser,* a privately published advertising magazine, similar in many ways to the modern, glossy, business promoters of today. An anonymous article appeared in that issue concerning Moses' activity as a political canvasser. Other than the knowledge that he was a voter for the House of Stanley, this is the only material I have discovered connecting him in any way to politics -

---

A PRESTON ASTRONOMER ELEVATED

Everyone who knew the late Moses Holden, the celebrated astronomer, was aware of his being a man of sterling worth, truly religious, and a strict teetotaller. But even men of high standing have been known, through perhaps mean strategy, to make a slip. So it was with Moses, but how was it brought about?

It was election time: Moses was a warm politician, and a capital canvasser. After a hard day's labour, the committee met at the Castle Hotel to compare notes and consume liquor. Moses was pressed to 'take something,' but declined.

A simple bottle of ginger beer was suggested as a cooler. Moses had no objection to this harmless beverage. A ginger beer bottle was filled with sparkling champagne, which Moses thought was very nice after the first taste, pronounced it very refreshing, and finished the bottle. To another bottle of the 'same sort of simple stuff' he had no objection.

This consumed, the venerable astronomer began to see stars not depicted on his orrery, and feel a motion of the globe not caused by an earthquake. His step became unsteady, and he had to be assisted home.

He was imposed upon; it was too bad, but it is nevertheless true.

It was often said that in the final third of his life Moses had become more and more difficult to deal with. Cantankerous and argumentative were just two of the adjectives used by those who knew him. I have certainly gained the impression that he didn't suffer fools gladly. However, a third of his life, or almost thirty years, takes us back to his days of founding the Institute. Following that he presented at least thirty-six courses of lectures in all parts of the old County of Lancashire, gathering glowing tributes along the way. So perhaps there was another side to the story?

As John Taylor commented in his work of 1885, *'Still justice and charity demand our praise and our tribute of respect for the work he did. According to his views of truth and righteousness, he was faithful and devoted in his service to his day and generation.'*

Epilogue

Following a talk that I gave about Moses Holden to the members of the appropriately named Halliwell Local Historical Society in Bolton in January 2014, I received a communication from the Society's secretary, Margaret Koppens. Whilst searching through a back copy of one of her local newspapers, her attention was drawn to an article about the man I had spoken about to her group.

She told me, "I already knew that there was an Observatory at Markland Hill from a photograph in our society's collection, but when I read the *Bolton Journal* and Guardian article, the name of the man jumped out at me, Mr. T. W. Holden, and I thought that there must be some connection with Moses Holden. Thanks to a colleague, Lois Dean, who researched the relationship for me, we now know that Thomas W. Holden, who presented the telescope to Bolton, was the grandson of Moses' elder brother, John."

I can confirm that by adding that this is the same man who was recorded as Master T. Holden, at Moses' funeral in 1864.

The article from the Bolton Journal and Guardian of the 17th October 1913 read as follows:

Despite the rain on Saturday afternoon many members and friends of the Bolton Field Naturalists' Society were present at the opening of the observatory which, through the assistance of the Parks Committee and friends who have given donations, the Astronomical sub-section of the Society has been able to erect in the market garden on Towncroft Lane. Mr. Atherton (Honorary Treasurer), in taking the chair, read a note of apology from Alderman Brooks on his inability to be present, and wishing the members success. Having expressed the pleasure of the Society that so many had turned out on such an unpleasant day, the chairman pointed out the difficulties that had been overcome in selecting the present site and erecting the building thereon, in which to house the four inch refractor telescope presented for the use of citizens in 1887 by the late Mr. T. W. Holden.

Replying to letters he had received suggesting difficulties of vibration (from motor traction and also local fogs) in carrying forward successful observations, he gave an invitation to the citizens to "come and see". He believed the site was the best possible, the terms for such were reasonable, the building was suitable and trusted the members would be successful in carrying out the work which Mr. Gibbs (who had kindly come over from Preston to open the observatory) would explain what was possible with the instrument therein.

He then asked Mr. George Joseph Gibbs, Fellow of the Royal Astronomical Society, Honorary Curator of the Deepdale Observatory in Preston, and for many years a leading light in the Preston Scientific Society, to unlock the door. After doing so, Mr. Gibbs explained to those present the possibilities of the telescope within the observatory for studying and mapping the features of the moon, the variable stars, Saturn and its rings, Jupiter and others, pointing out that much work might be done by amateurs in the study of astronomy, which had to be largely neglected by professional observers who were occupied with routine work. He also strongly advised the necessity of having suitable literature, and especially the necessary tables at hand, both before and whilst making observations. He advised the appointing of definite observers who should methodically use the instrument, and that it should not be left to what he might term, recreational observations. If the members got the best use possible out of the instrument and thoroughly exhausted its capacity, he felt sure that a wealthy town like Bolton would provide a similar, if not a better instrument to the one proud but poor Preston had provided out of Corporate funds as soon as it had been realised that useful astronomical work might be done in that town.

The instrument was next inspected by the members present, and Mr. Gibbs answered numerous inquiries as to its use and capacity, after which the thanks of the Society to Mr. Gibbs were moved by Mr. Holden, seconded by Mr. Parker and supported by Mr. Edward Bennis.

Appendices

ONE

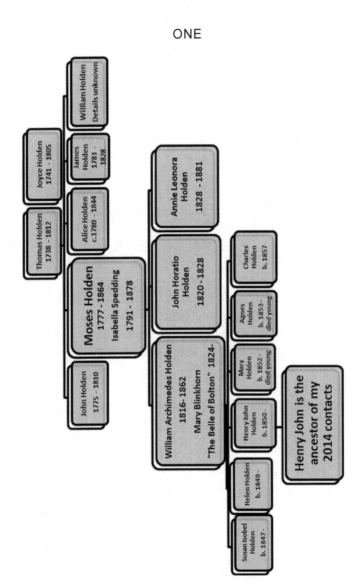

Thomas Holden 1738 - 1812

Joyce Holden 1741 - 1905

Alice Holden c.1780 - 1844

James Holden 1783 - 1828

William Holden Details unknown

John Holden 1775 - 1810

Moses Holden 1777 - 1864
Isabella Spedding 1791 - 1878

John Horatio Holden 1820 - 1828

Annie Leonora Holden 1828 - 1881

William Archimedes Holden 1816 - 1862
Mary Blinkhorn
"The Belle of Bolton" 1824-

Susan Isobel Holden b. 1847-

Helen Holden b. 1849-

Henry John Holden b. 1850-

Mary Holden b. 1852- died young

Agnes Holden b. 1853- died young

Charles Holden b. 1857

Henry John is the ancestor of my 2014 contacts

TWO

Paper to the Preston Literary and Philosophical Society.

"In compliance with your request I here compile my observations on the Comet of 1811. I have found that when the Comet made a near approach to any star, the star seemed nearer to the Comet than it really was. I have corrected most of these, and I flatter myself that what I present here will not be far from the truth of the Comet's apparent track.

On Wednesday September 4th about 8pm, I discovered this Comet in the north betwixt the hind legs of Ursa Major and over the back of Leo Minor, and in Right Ascension 160° and 39° of North Declination. I did not perceive much of the tail, but had every appearance of a star of the second magnitude.

On Thursday night the 5th, the Comet's tail was 4° in length, on the 6th it was considerably larger, but I did not perceive any considerable shifting until the 7th; but its' increasing in size all this while made me suppose that it was coming towards us. The Comet's train this night was 7° in length.

Wednesday September 11th I found the Comet to be in 160° of R.A. and 42° of N.D., and the Star Ψ of Ursa Major was in the centre of its tail, there appeared a convex and concave side of the Comet, the convex towards the west, and the concave towards the east. The head or nucleus seemed to be surrounded on that side towards the Sun with something of the appearance of a thick halo, that formed beneath the head at a distance from it, and seemed to wrap itself round to the tail behind both east and west, and form a thick, bright appearance both on the convex and concave sides, and leave a vacant place down the middle like that betwixt the head and that I term a 'halo'.

It appears to me through a nine feet refracting telescope (power 80), as if a transparent effluvia was emitted from the Comet equal on all sides, and impelled with some force to a

241

considerable distance from it – but meeting a superior force (and what this is deserves consideration), which fell back on itself where that impelled from the Comet meeting that repelled to it, there must be a place where both are in equilibrio, the effluvia still increasing forms a cloud where both forces meet.

The effluvia on the upper side (or on the side opposite the Sun) and the superior force impel or drive both one way, so that the concave and convex sides of the tail have all the effluvia from all the sides, except down the middle, much in the form of this:

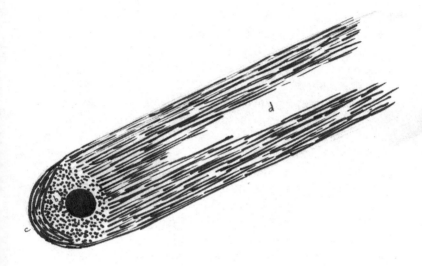

a. The direction the Sun is in.
b. The transparent effluvia impelled from the nucleus.
c. The halo which is formed turned east to west.
d. The vacant place.

But the middle is not entirely vacant but has the appearance that a flint glass bottle has, or a pair when seen through a microscope. The edges seem the densest.

September 17th. I found the Comet in 173° 47′ of R.A. and 45° 5′ of N.D., a star of the 3rd magnitude was in the centre of the tail, the Ψ of Ursa Major. The tail was 9° in length, and terminated 2° east of λ of the same constellation.

September 21st. The Comet's R.A. 180° 27′ and 46° 57′ of N.D.; the tail was between 9° and 10° in length. The east side of the tail terminated just clear of λ of Ursa Major, at the west of it, the west side 2° below δ of the same.

September 23rd. The Comet was in 184° 38′ of R.A. and 47° 40′ of N.D., the concave side of the tail was 1° 30′ above δ. This star was in the centre of the extremity of the tail, the concave side was below it, the length 12°.

September 26th. The light of the Moon took away the tail; the head looked like one of the stars of Ursa Major, and was beautifully centred in the five stars that compose the bow of the Great Bear betwixt λ and η.

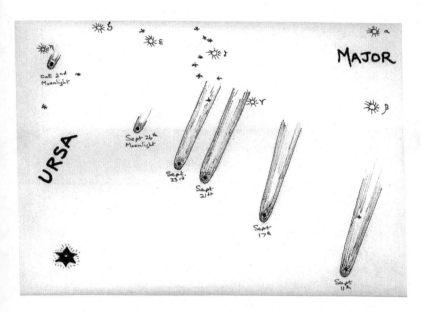

I am reliably informed that the bright star in the lower left of the picture is almost certainly Arcturus in the constellation of Bootes.

October 2nd. 9pm. The Comet was in 203° 50′ of R.A. and 49° 30′ N.D., 1° 10′ in R.A. short of η and 40′ less in Declination. The following were its positions whilst passing through Ursa Major.

On Friday the 4th October, I found the Comet to be in 208° 40′ of R.A. and 49° 45′ of N.D., and its tail pointed at a star of the 2nd magnitude, the α Draconis, and was then 15° 30′ long – this was the longest I saw the Comet's tail; it appears to me that the Comet was the nearest to the Earth either now or before I observed it again, not only from the length of the tail, but because the Comet at this time went over a larger apparent space than before or after. I had a view of the Comet this night with my nine foot refracting telescope, with an eye-piece that magnified eighty times. I had a beautiful sight of the circle I mentioned before. But I saw it more distinct and am more convinced of what I have before stated.

From the 4th to the 17th I had not my instruments with me although I made three observations I could not depend on them.

On the 17th I found the Comet had entered the constellation Hercules and was in 240° 50′ of R.A. and 44° 14′ of N.D.

On the 18th the Comet's R.A. was 243° 0′ and N.D. 43° 14′

On the 19th I found the Comet 10′ before δ of Hercules, a star of the 5th magnitude, the cloudy appearance of the Comet's head, sometimes nearly obscured that star, its R.A. 245° 10′ and N.D. 42° 15′.

October 22nd at 9 pm, I found the Comet to be in 151° 20′ R.A. and 39° 24′ of N.D. Its tail terminated betwixt λ and β Draconis.

October 25th. I find the Comet has passed the circle of perpetual apparition, and this was the first night that it would be set. This night it was 10° below π of Hercules, a star of the 3rd magnitude. R.A. 257° and N.D. 36° 54′. The ε of Hercules was in the centre of its tail, about 1° from the Comet's head. Time of observation 6 hr 30 min. PM.

October 27th. The Comet's R.A. was 260° 30′ and its N.D. 35° 25′; the Moon's light took away the appearance of the train.

October 30th. The Comet's apparent motion shortens every day. I found it in 265° 22′ of R.A. and 32° 20′ of N.D; the Moon gave too strong a light to suffer me to see the train.

November 1st. The Comet had just passed a star of the 5th magnitude, the λ of Hercules, about 20′ to the south-west of that star, and one degree perpendicular above ξ of the same constellation, R.A. 268° 10′ and N.D. 30° 10′.

On Monday night November 4th, I saw the Comet very brilliant, and the top part of the tail reached up to the 1 and 2 of Lyra, two stars of the 4th magnitude, on the concave side, but the convex went a little beyond β, a star of the 3rd magnitude. At 6.30pm the tail was perpendicular to the horizon and was 11° in length, R.A. 271° 50′ and N.D. 27° 13′.

November 7th. I had a fine view of the Comet; its tail reached to λ of Lyra, and had just passed that star, and was in length 9°. Its R.A. was 275° 20′ and N.D. 24° 25′.

November 10th. I found the Comet in 270° 15′ of R.A. and 22° 12′ of N.D.

November 11th I found it one degree of declination north of the star 110 of Hercules, Its R.A. 279° 20′ and N.D. 21° 25′.

On the 13th the tail terminated 2 degrees below β of Cygnus, a star of the 3rd magnitude, in the neck of the Swan and was then in 281° 15′ R.A. and 19° 45′ N.D.

November 16th. I found the Comet's tail to terminate at the mark 6 of Vulpecula in the nose of the Fox, and was then 8° in length. It had passed one branch of the Via lactea or Milky Way, and has now entered into that beautiful constellation Aquila. R.A. 283° 50′ N.D.17° 50′.

November 19th. Its R.A. 286° 10′. N.D. 15° 55′. The tail about 7° in length.

November 20th. Comet's R.A. 287° 0′. N.D. 15° 10′

November 23rd. Comet's R.A. 289° 20′ and N.D. 13° 20′. Here I gained a few minutes in N.D. not meeting with stars here, and taking my observations from some that the Comet passed near, so that the Comet seemed nearer than it really was, so it seemed further north.

On the 28th November, I got a good observation and found the Comet in 292° 50′ of R.A. and 10° 50′ of N.D.

November 30th. I found the Comet 17′ below the λ of Aquila, and 294° 0′ R.A. and 9° 50′ of N.D.

December 2nd. The Comet is now 38′ above α of Aquila, and in 295° 25′ of R.A. and 8° 53′ N.D.

On the 4th the Comet had passed by the ξ of Aquila and was 296° 28′ R.A. and 7° 57′ N.D.

Day	R.A.	N.D.
December 5th	296° 50′	7° 30′
9th	298° 55′	6° 5′
10th	299° 20′	5° 38′
16th	301° 32′	3° 55′
19th	302° 30′	3° 10′

December 30th. This night I just perceived the Comet, but the Moon rose before I could take a proper observation; this will not be far from the truth R.A. 308°, and 1° S.D.

The inclination of the orbit of this Comet will be about 65° 30′ to the Ecliptic.

THREE

Nb. A recapitulatory lecture was given where a fourth date is shown. They were a compilation of the three main lectures, usually delivered for half the usual admission charge. The operative classes were usually allowed entrance for 6d (2½p). There were probably many more instances than are indicated in these records.

1806	Bolton Dates unknown
1815	Preston 7th 10th and 11th April
1815	Lancaster 26th 28th April, and 1st May
1815	Chester July
1815	Liverpool 11th 14th and 15th August
1816	Leeds February
1817	Gainsborough Dates unknown
1817	Nottingham 1st 2nd and 3rd September
1818	Preston 16th 17th 19th plus 23rd March
1818	Manchester Minor Theatre End of June
1818	Nantwich November
1823	September 18th *Attendance at Horncastle Methodist Circuit Annual Meeting*
1824	*Northallerton – Sermons*
1824	York September
1825	Carlisle November
1826	Whitehaven April
1826	Liverpool 20th 21st 22nd November

1827	Preston 26th 27th 28th March, and 2nd April
1828	Preston (Optics)Early April
1829	January 18th *Holbeck – Sermons*
1829	Preston (Institute for the Diffusion of Knowledge) 16th April.
1830	*Holbeck – Sermons*
1830	Preston 26th 27th 29th April
1831	Lancaster 25th 26th 27th April
1831	Kendal 16th 17th 18th May
1833	Preston April
1833	Chorley Dates not known
1833	Wigan Dates not known
1833	Bolton Late October
1833	Theatre Royal, Manchester December
1833	Oldham Dates not known
1833	Stockport Dates not known
1834	Queen Theatre, Manchester mid-January
1835	Liverpool w/c 2nd and 9th November. (two courses)
1836	Preston w/c 18th April
1837	Lancaster 22nd 23rd and 24th May
1837	Liverpool w/c 5th November
1837	Warrington End of November
1838	Rochdale October (two courses)
1838	Halifax November
1839	Over Darwen25th 26th and 27th September
1839	Blackburn21st 23rd and 24th October; 29th 30th and 31st October

1840	Manchester 23rd 24th 25th 30th 31st March and 1st April (Two Courses)
1841	Lancaster August
1842	Preston End of April
1844	Liverpool, Liver Theatre 20th 21st 22nd 25th 26th 27th 28th 29th and 30th March [Three Courses]
1844	Warrington 22nd 23rd and 24th April
1845	Preston 15th 16th and 18th September
1849	Preston 24th 25th 27th September, and 1st October
1851	Ormskirk February
1852	Preston 11th 12th 14th and 18th October

FOUR

Optics
M. Holden
16th April 1829
At the Cockpit.

The lecture began by taking a short review of the subjects treated in his former lecture, viz. Light, Colours, and the Human Eye. On the subject of colours he observed that he had before omitted to notice one circumstance which might be of importance to some of his hearers, as there might be some among them who were painters[1]

He would ask why did not modern paintings, equal in point of colouring those of the ancients. He would tell them. It was because the moderns used the pallet knife, and that corrodes colours. They who wanted paint to tell a tale a hundred years hence must not use the pallet knife. Let them use a piece of ivory or glass, and this would answer every purpose.

With regard to the human eye too, there were several particulars connected with that subject which he omitted to notice before, and particularly a simple remedy for the cure of inflamed eyes, adding that he had been unwell at the time and that his memory had failed him.

He continued by saying that he did not wish to spoil the trade of the doctors, but he saw no harm in telling his audience of such a simple remedy – and was nothing more than a steaming of the eyes. *'Set a kettle on the fire, and it is an odd kind of family that does not use a kettle. Don't let the water be higher than the end of the spout inside, for if you do, the steam would not come out. Blow the fire briskly, and when the steam comes out of the kettle, apply a tube of brown paper and let the steam flow through this upon the eye. From this simple remedy great relief may be had. Even constitutional sore eyes may be benefitted by this means. If, however, the eyelids be inflamed, steam must not be applied to them for it will injure them'.*

Pers. Comm. Moses wrote about using this method to treat individuals whilst on his missionary tour in 1811 – 1812.

The present lecture was of other things connected with vision, viz. of the assistance afforded to it by glasses; every instrument for the purpose of assisting vision whether a telescope or a microscope must be made of glass or of something transparent.

The lecturer then described the process of grinding and polishing magnifying glasses and the mode of cutting glass with shears made purposely. He observed that glass might be cut with common pliers.

He then exhibited a compound microscope and gave an account of its power, construction and invention. The microscope he had out before him would magnify seventy millions with a power of 70 millions – through it *'the little creatures in children's heads appeared as large as a human body'*, and with the highest power the lees in vinegar and in sour paste appear a yard long.

The first microscope was made in the year 1619 by Lewino; the merit of the invention was claimed about nine years afterwards by a Dutchman, and about twenty-seven years after the invention by Foulenoy. *[NB. We must presume that this information was believed to be authentic at the time – 1829 – but modern knowledge doesn't mention any names that resemble the ones given here].*It was rather singular that it was not until nine years after the telescope was invented that the microscope was made, they being so similar in principle, so much so, indeed, that he (the lecturer), sometimes made the former serve the purpose of the latter.

Lewino, in his microscope, used a single lens only – and with this he perceived the smallest animalcula. He must have had good eyes, and must have strained them by looking through his single lens microscope.

The compound microscope does not injure the eye. The lecturer had used one since he was seven years of age, and his eye was not at all impaired by using it.

Speaking of animalcule, the lecturer remarked, that in certain seasons of the year they may be found in everything, except in pure water. It had been said that they were to be found even in pure water, but this he denied. He had examined it with a power of many millions and found nothing alive in it. However, pure water, after being exposed for a short time to the air will have animalcule in it. He had also examined vinegar through the microscope; in some he found no animalcula, while in others he found many, and in a single half gill of some he had discovered as many as the inhabitants in this kingdom. The animalcula in vinegar are of a white colour.

The lecturer observed that many cutaneous diseases are caused by nothing but animalcule in the skin. In chilblains, for instance, he had examined the part frost-bitten. On one occasion he looked at a drop of matter from a chilblain through the microscope and perceived no less than three mites in it, each having six legs and covered with hairs. These mites were of a very hardy kind and must be of a very destructive nature. Of their hardihood he had ample proof from experiments. He had first tried if they could live in water, which they could do. He then tried them in turpentine and this they also braved – and there are few such insects which can withstand it - flies will die in turpentine.

He then took some tincture of myrrh, but these 'little gentlemen' waded through that, too, and 'seemed to say, we are the conquerors, we can have this also'. He at last tried them in sulphuric acid, and they immediately died. Let a person affected by such disorder wash the parts affected in sulphuric acid diluted (and which will do them no harm) and they will derive a benefit. The lecturer knew a person in Oxfordshire effectually cured by such application.

An intense itching of the eye is often felt by some. This is caused by nothing more than animalcule, but he would not have anyone rub his eye with sulphuric acid. If he did he would stand in need of a large dancing room – Marshal's Cereal is very good for itching eyes.

It was owing to optics that these discoveries had been made. Speaking of the microscope, he said that all nature was replete with wonders through this instrument, and nothing presents a more beautiful appearance through it than the duck weed – he would not wish to behold a more beautiful menagerie than weed presented through the microscope. Some animalcula had roots.

The water in Fen countries such as Lincolnshire and Cambridgeshire had many animalcula in it. Some time ago he was in the latter county. It was in Christmas week, and the season frosty. Being in the habit of taking a glass of water before he went to bed (for it made him sleep more soundly) he one evening called for a glass. The water he fancied smelled very strongly of fish. He took his microscope through which he viewed a single drop, and in that drop he perceived at least ten thousand living creatures. His audience might suppose that he was not going to swallow them in such multitudes. He changed his beverage to ale. It, however, brought on a disease (the bile) which he feared he would never get rid of. Intermittent fevers he observed were frequently occasioned (by drinking foul water).

He then described the solar microscope and said that he had lately made a most interesting discovery in producing a variety of most beautiful colours – equal in point of brilliance to those of the rainbow, and this by the use of the microscope and the phantasmagoria[2].

He concluded his lecture by a description of the various kinds of glasses – telescopes – perspective, and day and night glasses, their construction, power, mode of acting and use; and he illustrated his observations by diagrams which he showed with great promptness and ingenuity.

[1] The writer here inserted a footnote, observing that the lecturer was mistaken about the omission, for he had mentioned it in the previous lecture.

[2] Phantasmagoria: a name invented for an exhibition of optical illusions produced chiefly by means of a 'magic lantern'.

Anonymous reporter.

FIVE

Memorial to those who were founders of the Institution:

Robert Ascroft, Attorney; John Atherton, Mechanic; Thomas Barker, Draper; Adam Booth, Mechanic; Lawrence Booth, Mechanic; George Cowperthwaite, Sedan Carrier; Richard Dunn, Mechanic; John Gilbertson, Surgeon; Moses Holden, Gentleman; John Johnson, Tailor; Joseph Livesey, Cheesemonger; Josiah Lyon, Joiner; Edward Makin, Cotton Manufacturer; Francis Nelson, Mechanic; Robert Norris, Gardener; Thomas Pritt, Engraver; George Riley, Gentleman; John Robinson, Overlooker; James Tomlinson, Shopkeeper; William Toulmin, Coal Dealer; Peter Walmsley, Joiner; Michael Whaling, Twist Maker; Francis Wilkinson, Tailor.

First Committee of the Institute for the Diffusion of Knowledge:

Thomas Batty Addison, Esq. President; Joseph Livesey, Treasurer; Robert Ascroft, Secretary;

John Gilbertson; Adam Booth; Peter Walmsley; George Hodgson; Moses Holden; George Riley; William Toulmin; Thomas Pritt; Edward Makin; Thomas Barker; Francis Nelson; John Robinson; John Johnson; Michael Whaling; Robert Norris; Josiah Lyon; James Tomlinson; Richard Dunn; Lawrence Booth; Francis Wilkinson; John Atherton.

SIX

The following poems are taken from the 'Life and Poems' of Henry Anderton, a poet from Walton-le-dale, who was a particular follower of Moses Holden, and who would walk from Walton-le-dale to the chapel in Vauxhall Road, Preston, to hear him preach. At the time of the secession of the Protestant Methodists denomination from the Wesleyan Methodist body, Moses' sympathies lay with the secessionists, and with them he assisted in the erection of the Vauxhall Road Chapel in Avenham. He preached there until the time he decided to become a member of the Church of England, and to attend Christ Church.

The first of the poems takes the form of an imaginary conversation with his mother, who, it would seem, was less enthusiastic about visits to Vauxhall Road than he was.

Henry:	Dear Mother, shall I go to Preston?
	Answer me this important question.
Mother:	You rascal you, pray let me know
	Why you so urgent are to go.
Henry:	I'll tell you if you'll cease your scolding:
	I want to hear old Moses Holden.
	Mother, don't cast my spirits down,
	But let me have a walk to th'town.
Mother:	All things are proper in their seasons:
	You want to go, but what's your reasons?
Henry:	Nay! Do not with my feelings grapple,
	But let me go to Vauxhall Chapel.

The following was written on a Good Friday:

> This day to me could not have been worse,
> Instead of a blessing it has been a curse. But stay!
> Before thou begin'st to chatter,   And endeavour to
> cut more short the matter: In spite of what earth to
> my aim opposes, Let me go to Preston, to hear old
> Moses.

On another occasion he wrote:

> Tonight old Moses will, in Vauxhall Road,
> Show unto sinful men the way to God.
> Dear mother, shall I go? Grant my request!
> Whether I'm right or wrong you know the best.
> I want to go to Vauxhall Road tonight;
> Be as it seemeth good unto your sight.
> But you will, perhaps, both think and say, "Thou'rt
> hollow;
> Thou hast some other end or aim to follow." Believe
> me, mother, I have not; and so
>
> Speak out the cheerful words, "Son, thou must go."

Henry Anderton seems to have been greatly
impressed with the preachers in the Vauxhall
Chapel.

> Away with the trash of your college-taught
> preaching! Nor prate in the hearing of me or of
> mine;
> Let me have the truths of the Gospel, heart
> reaching, Though blunders grammatical spring in
> each line.
> Away with that minion of fame and of glory,

That man-pleasing preacher, conceited and vain!
I'd choose the poor rustic, whose plain, simple story
Is pointing to Jesus, the Lamb that was slain.
Away, child of lucre! Thou holdest a station
A station not far from the precincts of hell   Who
makest the work of eternal salvation   A system of
traffic, to buy and to sell.
Ye Vauxhall Road preachers, sincerely I love you;
Your labours, I trust, will be own'd of the Lord:
For what but the love of Jehovah could move you
To preach, without money, the life giving word.
Whene'er from the body the Saviour shall free you,
And raise unto glory your justified souls,
There, cloth'd in His likeness, I hope I shall see you
Among the redeem'd while eternity rolls.

To Mr. Holden, of Preston, on the Baptism,
of the Prince of Wales.

The morning dawned on which to give
A name to Britain's heir;
To celebrate the great event,
The good and true prepare.
And what she owes that regal house   "Our Village" can't
forget;
For Walton is a loyal place, And glories in the debt.
Hark! Through her lanes and alleys, now,
This prayer-like shout prevails, "God Save the Queen and
Albert, And the Royal Prince of Wales."
And on that day our scholars, too,
Made up a gallant show;
With flags and banners, manfully

They toddled through the snow.
Nor was good "inside plenishing"
For their young ribs forgot;
For a cartload of buns they had, And coffee piping hot!
And never yet a blither crew
On British ground was seen,
To bid a Prince right welcome,   And to sing God save the Queen!"
The aid was sought, and soon forgot
Was all about the stars,
How Saturn calls on Jupiter
And Venus plays with Mars.
Discarded were the lens and tube,
Down came thy magic glass;   And when before the children's gaze
Thy mimic ghosts did pass,
Pure rapture filled thy kindling heart,
With watching the surprise
Which stretched the youngsters' little mouths,
And strained their little eyes.
We love thee, Moses Holden; for
Old Brunswick's Royal line
Could never boast a faith more true, A heart more warm than thine,
Respected for thy loyalty,
As honoured in thy fame;
And in this verse thy Walton friends,
With one consent, exclaim -
(Thou Whigs may hate thee for this cause, and Radicals revile)'
Behold an Englishman indeed,   In whom there is no guile.

Moses Holden's Recovery from Sickness.

It's a dark and awfully stormy night,
What dreadful forebodings the traveller afright;
No light, save the lightning's lurid glare
Shooting across the sulphurous air;
Bewildered he falls, and yields to despair.
At this desperate crisis, as from a shroud,
The moon appears from behind a cloud;
Her beams on the face of the traveller alight;   His bosom is
fired with hope at the sight;
His way he resumes with unwonted delight.
Just so were we; no rest could we find
While a stroke of affliction our preacher confined;
But now he's restored, he rekindles our joys;
The air we will rend with a thanksgiving noise,
The heavens shall echo the sound of our voice.

The following was written on the occasion of Moses
Holden leaving Vauxhall Road Chapel:

"ICHABOD"

Vauxhall! Thou must not hear him more;
His labours in thy wall are o'er;
Vauxhall, the fatal die is cast,
The dreaded Rubicon is past;
No more thy fame will spread abroad,
Thy name is changed to *"Ichabod,"*
Thy glory is departed.
Vauxhall! Thy elders are ingrate,
Factious compounds of scorn and hate.
I saw them at the lovefeast, while

They should have prayed, with fiendish smile,
They scowled upon the man of God: These men have
named thee *"Ichabod."*
Thy glory is departed.
Vauxhall! Thou hast thy preacher lost -
Of thee the glory and the boast;
Chased by the ingratitude of those
Who, seeming friends, were mortal foes.
Forgive, convert them, gracious God!
They've named thy Temple *"Ichabod,"* *They glory is*
departed.

SEVEN

An Ideal Orrery

An early nineteenth century explanation.

Conceive the Sun represented by a globe two feet in diameter; at 82 feet distance, put down a grain of mustard seed, and you have the size and place of the planet Mercury, that bright silvery point which is generally enveloped in the solar rays. At a distance of 142 feet lay down a pea; it will be a similitude of Venus, or dazzling evening and morning star.

215 feet from the central globe, place another pea, only imperceptibly larger, that is man's world, (once the centre of the Universe!) the theatre of our terrestrial destinies, the birth-place of most of our thoughts!

Mars is smaller still, a good pin's head being his proper representative, at the distance of 327 feet; the four smaller planets, Vesta, Juno, Cores and Pallas, seen as the least possible grains of sand, about 500 feet from the Sun; Jupiter, as a middle-sized orange, distance of about a quarter of a mile; Saturn, with his ring, a larger orange at the distance of two-fifths of a mile; and the far Uranus dwindles into a cherry, moving in a circle ¾ mile in radius.

Such is the system of which our puny earth was once accepted the chief constituent; a system whose real or absolute dimensions are stupendous, as may be gathered from the Sun himself, the glorious globe around which these orbs obediently circle; which has a diameter nearly four times larger than the immense interval which separates the Moon from the Earth

EIGHT

(A) List of Sponsors of 1852 Farewell Lectures.

1852 SPONSORS

Abraham J.
Addison T.B.
Ainsworth H.T.W
Ainsworth T
Allen G.F.
Anderton R.
Armstrong Rev. A.
Arrowsmith R.
Ascroft R.
Backhouse J.D.
Bairstow J.
Birley Mrs.
Booth J.B.
Broughton Dr.
Cartwright S.
Carus Mrs
Catterall J.
Catterall Paul
Catterall Peter
Chambers J.
Clarke Rev. T.
Clough T.
Corless Mr.
Croft B.

Dawson H.
Dickson W.
Dodd T.
Dracow F.
Duckett Mr.
Edmondson J.
Farish Rev. W.M.
Fisher J.
Gardner E.T.
German J.
German Mrs.
Gilbertson J.G.
Girley C.
Glass J.
Glaudretto J.M.
Goodwin Mrs.
Gordair J.
Gorst E.
Gorst Mrs.
Gorst Miss
Gorst Miss M.E.
Graves H.
Grower Mrs. G.
Grower Mrs J.

Grower R.
Grundy Mr.
Grundy J.
Grundy T.
Haines J.
Harris Rev. R.
Haslam Mrs. J.
Haslam Miss
Haydock E.
Henry Miss.
Higgins J.F.
Hincksmann T.C.
Hogg J.
Hollins E.
Hopkins R.M.
Horrocks J.
Horrocks Mrs.
Houlker T.
Humber J.
Jacson C.R.
Jennings H.
Jones Rev. W.P.
Kitton Rev. H.
Leach T.
Leay Mrs.
Loundes T.
Lowe R.
McGuffogg Mr.
Marsh J.
Meek Messrs.
Melling P.M.
Miller T.
Monk T.
Moore F.

Myers J.J.
Myers M.
Naylor J.
Newsham R.
Norris J.H.
Norris T.
Oxendale T.
Paley J.
Palmer R.
Park R. (Withnell)
Parker R.
Parker R.J.
Parr Rev, J.O.
Pateson Mr.
Pedder E.
Platt Mr.
Rodgett M.
Seed J.
Segar R
Shawe H.
Shuttleworth T.S.
Smethurst J.
Smith Rev. H.H.
Spence L.
Stavert Dr.
Stevenson J.
Story Miss.
Swainson C.
Swainson J.
Threlfall R.
Threlfall T.
Underwood T.
Walker J.
Walmesley T.

Walton H.C.
Whitehead J.
Wilcockson I.
Wilson E.
Wilson S.
Winstanley G.
Winstanley J.
Winstanley Miss.
Woodfords

(B) Accounts of the 1852 Farewell lectures at the Preston Theatre Royal.

(C)

1st night. 11th October:

		£	s	d
Money		4.	1.	6
Box Tickets		25	0	0
Pit & Gallery Tickets		3	8	6
Grammar School	49 pupils	3	5	4
Miss Ogle	19 pupils	1	5	4
Mrs. Buchanan	15 pupils	1	0	0
Miss Romanis	12 pupils	1	6	0
Miss Hickney	15 pupils	1	0	0
Miss Turton	6 pupils		8	0
Rev. H. Power	5 pupils		6	8
		40.	11.	4

2nd night 12th October:

		£	s	d
Money & J. Sidgreaves (sic)		4	1	0
Box tickets		19	5	4
Pit & Gallery tickets		3	16	4
Grammar School	(49)	3	5	4
Miss Ogle	(16)	1	1	4
Miss Buchanan	(15)	1	0	0
Miss Romanis	(12)		16	0
Miss Hickney	(15)	1	0	0
Miss Turton	(5)		6	8
Rev. H. Power	(5)		6	8
T. Lowndes		2	0	0
		36.	18.	8

3rd night 14th October:

Money	6	2	6
Box tickets	22	18	0
Pit & Gallery tickets	4	1	0
Grammar School	2	2	8
Miss Ogle	1	8	4
Miss Romanis		16	0
Miss Hickney	1	1	6
Miss Turton		6	8
T. Lowndes	1	0	0
T. Parker	2	0	0
	41	6	8

3rd night 14th October:

Money	6	2	6
Box tickets	22	18	0
Pit & Gallery tickets	4	1	0
Grammar School	2	2	8
Miss Ogle		18	4
Miss Romanis		16	0
Miss Hickney	1	1	6
Miss Turton		6	8
T. Lowndes	1	0	0
T. Parker	2	0	0
	41	6	8

4th night 18th October: £7 14 0

Expenses

	£	s	d
Rent of Theatre	5	0	0
Printing	2	11	6
Advertising	2	0	6
Oil and candles		5	10
Gas		4	6
Posting and delivering		12	0
		7	0
Envelopes and postage stamps		5	6
Joiners, use of wood &c.		9	0
Jane, cleaning		4	10
David Baxter		7	0
Thomas Harrison		9	0
William for helping	6	13	0
Annie for helping	2	0	0
	21	9	8

Total receipts	126	10	8
Total expenses	21	9	8
Nett profit	£105	1	0